航空维修工程英语

胡静 ◎ 主编
张铁纯 陈聪 ◎ 副主编

清华大学出版社
北京

版权所有，侵权必究。举报：010-62782989，beiqinquan@tup.tsinghua.edu.cn。

图书在版编目（CIP）数据

航空维修工程英语 / 胡静主编. -- 北京：清华大学出版社，2025.1. -- ISBN 978-7-302-67789-5

Ⅰ. V267

中国国家版本馆 CIP 数据核字第 20256B531W 号

责任编辑：王　欣
封面设计：何凤霞
责任校对：欧　洋
责任印制：沈　露

出版发行：清华大学出版社
网　　址：https://www.tup.com.cn,https://www.wqxuetang.com
地　　址：北京清华大学学研大厦 A 座　　邮　　编：100084
社 总 机：010-83470000　　邮　　购：010-62786544
投稿与读者服务：010-62776969，c-service@tup.tsinghua.edu.cn
质量反馈：010-62772015，zhiliang@tup.tsinghua.edu.cn
印 装 者：小森印刷霸州有限公司
经　　销：全国新华书店
开　　本：185mm×260mm　　印　张：16.75　　字　数：403 千字
版　　次：2025 年 1 月第 1 版　　　　　　　印　次：2025 年 1 月第 1 次印刷
定　　价：62.00 元

产品编号：107644-01

前言

为适应现代飞行器设计与航空维修人才培养的需要，本教材在编写过程中紧跟民航应用技术前沿，选材精练实用，突出课程"能用，实用"的培养目标，适合航空器设计、制造和维修以及航空机械电子专业本科生的专业英语学习，也可为从事航空器设计、制造和维修工作的工程技术人员提供参考。

本书以教学单元（UNIT）作为基本编写模块，构建涵盖现代民航工作领域的科技英语宏观框架，具体划分为两个层次：第一层为航空基本知识，包括飞机发展、飞行原理、飞机设计、飞机材料、飞机无损检测及腐蚀控制；第二层为现代民机典型系统，包括电子电气系统、机械系统、发动机系统。根据教学目标，每个教学单元分为三个知识模块：精读课文及专业词汇、阅读材料、维修工程英语专题，旨在提升学生的科技英语应用能力，书中给出的词汇也是飞机设计与维修工作中高频词汇。本书素材多选自国外经典民航科技出版物，比如美国 FAA 的 *Aviation Maintenance Technician Handbook* 等，文章语法、句式、选词符合航空维修简化工程英语要求；讲述的航空专业知识选材新颖，紧跟民航一线生产实践，例如燃油箱惰化系统、空调加湿系统等，展现民航技术发展动态。另外，在教材每个单元后设置技术专题，紧紧围绕科技英语特点、翻译技巧及航空维修工程英语的实际应用，真正做到学以致用。

本书由中国民航大学胡静教授担任主编并统稿，张铁纯、陈聪担任副主编。书中正文 UNIT 1、UNIT 10、UNIT 11、UNIT 18 由张铁纯编写；UNIT 2、UNIT 14 由谭娜编写；UNIT 3、UNIT 9、UNIT 15、UNIT 16 由胡静编写；UNIT 4 和 UNIT 5 由张国尚编写；UNIT 6、UNIT 7、UNIT 13 由陈聪编写；UNIT 8、UNIT 19、UNIT 20 由牟清源编写；UNIT 12、UNIT 17 由东方航空技术有限公司崔相国编写；书中专题内容由中国民航大学胡静、张铁纯、牟清源、梁大敏和东方航空技术有限公司云南分公司李茂云编写，南方航空公司沈阳分公司汪涛也参与了本书的部分编写工作。

中国民用航空飞行学院蒋陵平教授和空客公司代表刘晓阳对本书初稿进行了审阅，并提出诸多宝贵意见，对他们表示由衷感谢！中国民航大学 20 级硕士研究生张涵哲、21 级硕士研究生商雪晨承担了本书的图形、图像处理工作，在此一并表示感谢！

中国民航大学张艳玲教授、中国民用航空飞行学院刘峰教授对本书进行了终稿审校，在此代表编写组对两位教授致以深深谢意。

由于编写时间仓促及编者水平有限，书中错误和疏漏之处在所难免，敬请各位专家和读者予以指出，以便再版时加以纠正。

编 者

2024 年 5 月于天津

目 录

UNIT 1　INTRODUCTION　1
　　TOPIC 1：科技英语特点　9
UNIT 2　PRINCIPLE OF FLIGHT　13
　　TOPIC 2：专业英语翻译标准与技巧　24
UNIT 3　AIRCRAFT DESIGN BASIS　28
　　TOPIC 3：航空维修英语特点　37
UNIT 4　TYPES AND APPLICATIONS OF AEROSPACE MATERIALS　39
　　TOPIC 4：航空维修英语特点（续）　47
UNIT 5　NDT AND CORROSION OF AEROSPACE MATERIALS　50
　　TOPIC 5：简化技术英语　59
UNIT 6　ELECTRICAL SYSTEM　63
　　TOPIC 6：简化技术英语（续）　72
UNIT 7　AVIONICS SYSTEM　75
　　TOPIC 7：常用数学符号和公式读法　86
UNIT 8　AIR CONDITIONING SYSTEM　88
　　TOPIC 8：单位及单位换算　100
UNIT 9　FLIGHT CONTROL SYSTEM　102
　　TOPIC 9：字母与数字特殊读法　113
UNIT 10　AIRCRAFT FUEL SYSTEM　115
　　TOPIC 10：ATA100 规范　124
UNIT 11　HYDRAULIC SYSTEM　127
　　TOPIC 11：飞机站位及区域划分　137
UNIT 12　LANDING GEAR SYSTEM　141
　　TOPIC 12：机务放行英语口语　154
UNIT 13　LIGHTING SYSTEM　159
　　TOPIC 13：飞机主要维修工具　168
UNIT 14　OXYGEN SYSTEM　169
　　TOPIC 14：材料相关专业词汇　179
UNIT 15　FIRE PROTECTION SYSTEM　181
　　TOPIC 15：飞机缺陷常用术语　193

UNIT 16	**PNEUMATIC SYSTEM**	197
	TOPIC 16：ICAO 简介	204
UNIT 17	**ICE AND RAIN PROTECTION**	206
	TOPIC 17：维修工作签字常用句	217
UNIT 18	**INERTING GAS SYSTEM**	219
	TOPIC 18：工程指令编写规范	228
UNIT 19	**ENGINE CONSTRUCTION AND PRINCIPLE**	232
	TOPIC 19：科技论文写作规范	242
UNIT 20	**ENGINE FUEL AND CONTROL SYSTEM**	245
	TOPIC 20：英文电子邮件写作规范	256
参考文献		259

UNIT 1

INTRODUCTION

[1] A brief history of **development of aircraft**: The first free flight in human history took place on November 21, 1783, when the Montgolfier brothers successfully launched a hot air balloon. Just two years later, two explorers, Frenchman Jean-Pierre Blanchard and American John Jeffries, made history by crossing the English Channel in a balloon, completing the journey in two and a half hours. Later, airships emerged. The biggest difference between airships and balloons is that the former have certain maneuverability and directionality, while the latter are subject to the effect of wind.

[2] At the beginning of the twentieth century, two brothers from the United States made great contributions to the advancement of aviation. They were the Wright brothers. At that time, most people thought that it was impossible for an airplane to fly on its own power, but the Wright brothers did not believe this. From 1900 to 1902, the brothers carried out more than 1,000 glider test flights, and finally produced the "Flyer I" in 1903, and the test flight was successful. This was a great progress in the history of aircraft development.

[3] From 1927 to 1932, progress was made in the development of cockpit instruments and pilot equipments, and gyro technology was applied to flight instruments. The rotating flywheel, mounted on the gimbal, maintained spatial orientation and became foundational for various navigation tools, enabling pilots to fly safely in darkness, rain and snow. During this period, the artificial horizon instrument came to be installed on the aircraft, which could indicate the flight attitude of the aircraft to the pilot; the gyromagnetic compass indicator, with degrees marked on the compass; and the geomagnetic induction compass, which was not affected by the large amount of iron things often carried on the aircraft, nor by the vibration and the earth's magnetic field. These instruments, with high sensitivity, include altimeters which could measure altitude more than 30 meters above the ground, turn sideslip indicators that display aircraft's turning angle speed, and radio beacons which indicate air routes, are all used to guide pilots through unfamiliar terrain.

[4] World War II greatly stimulated the development of aircraft, and aircraft technology achieved its second leap. A new power aircraft-jet aircraft came onto the stage of history. On July 18, 1942, German aeronautical engineers reached a milestone in aviation history

when the twin jet powered Messerschmitt Me 262 made its first flight under jet power.

[5] **Boeing aircraft**: Boeing series aircraft is a very successful civil transport aircraft product series owned by Boeing Company of the United States. B737 series aircraft is a kind of short and medium range double engine jetliner. B737 has been on sale for more than 60 years since it was put into operation.

[6] The B737, including B737-200/-300/-400/-500, is based on a key Boeing philosophy of delivering added value to airlines with reliability, simplicity and reduced operating and maintenance costs. The B737-6001-700/-800/-900ER models continue the B737's pre-eminence as the world's most popular and reliable commercial jetliner. B737MAX is the newest model of the B737 family.

[7] **Airbus aircraft**: A320 series is a double engine, medium and short range, 150 seat class passenger aircraft, and developed by European Airbus Industry Company. The fuselage adopts the common semi-monocoque shell structure, and the cross-section is circular. The main components are made of high strength aluminum alloy, steel and titanium alloy. The A320 family are low-wing cantilever monoplanes with a conventional tail unit with a single vertical stabilizer and rudder. Two turbofan engines are installed under the wings.

[8] The A320 family is based on a common aircraft type with the same wide cabin cross-section, which is available in four fuselage lengths: the A318, A319, A320 and A321 —all of which share the same systems, cockpits, operating and maintenance procedures, and pilot type rating. They cover the market from 100 to 220 seats and currently fly on ranges of up to 3,700 nautical miles. The family's cornerstone aircraft is the A320, which accommodates 150 passengers in a typical two-class arrangement, and up to 180 with high-density seating. The stretched-fuselage A321 version seats 185 passengers in the two-class layout, and up to 220 for a high-density cabin. The shorter-fuselage A319 has a 124-passenger capacity in the two-class configuration, and up to 156 in high-density, while the smallest Airbus' A318 seats 107 passengers in the two-class cabin and 132 with high-density seating. A320NEO is the newest model of the A320 family.

[9] **Cockpit**: Cockpit is the flight compartment in which the pilot controls the aircraft, usually located in the front of an aircraft. In addition to some of the early aircraft, most of today's aircraft have a closed cockpit. The cockpit of an aircraft is generally equipped with various flight instruments and aircraft control systems (Fig.1-1). In most civil aviation aircraft, a door is used to separate the cockpit and passenger cabin. After the terrorist attack on September 11, 2001, most airlines secured the door to prevent hijacking and other accidents.

[10] Until recently, most general aviation (GA) aircraft were equipped with individual instruments utilized collectively to safely operate and maneuver the aircraft. With the release of the electronic flight display (EFD) system, conventional instruments have been replaced by multiple liquid crystal display (LCD) screens. The first screen is installed in front of the left seat pilot position and is referred to as the primary flight

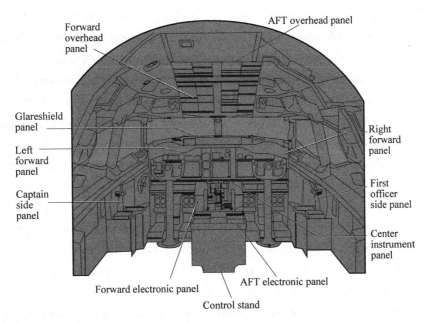

Fig. 1-1 Cockpit

display (PFD). The second screen, positioned approximately in the center of the instrument panel, is referred to as the multi-function display (MFD). These two screens de-clutter instrument panels while increasing safety. This has been accomplished through the utilization of solid state instruments which have a failure rate far less than those of conventional analog instrumentation. With today's improvements in avionics and the introduction of EFDs, pilots at any level of experience need an astute knowledge of the onboard flight control systems as well as an understanding of how automation melds with Aeronautical Decision-Making (ADM). Whether an aircraft has analog or digital ("glass") instruments, the instrumentation falls into three different categories: performance, control, and navigation.

[11] **Performance instruments**: The performance instruments directly reflect the performance the aircraft is achieving. The speed of the aircraft can be referenced on the airspeed indicator. The altitude can be referenced on the altimeter. The aircraft's climb performance can be determined by referencing the vertical speed indicator (VSI). Other performance instruments available are the heading indicator, angle of attack indicator, and the slip-skid indicator.

[12] **Control instruments**: The control instruments display immediate attitude and power changes, and are calibrated to permit adjustments in precise increments. The instrument for attitude display is the attitude indicator. The control instruments do not indicate aircraft speed or altitude. In order to determine these variables and others, a pilot must reference the performance instruments.

[13] **Navigation instruments**: The navigation instruments indicate the position of the

aircraft in relation to a selected navigation facility or fix. This group of instruments includes various types of course indicators, range indicators, glideslope indicators, and bearing pointers. Newer aircraft with more technologically advanced instrumentation provides blended information, giving the pilot more accurate positional information. Navigation instruments are comprised of indicators that display global positioning system (GPS), very high frequency (VHF), omni-directional radio range (VOR), nondirectional beacon (NDB), and instrument landing system (ILS) information. They also provide pilotage information so the aircraft can be maneuvered to keep it on a predetermined path. The pilotage information can be in either two or three dimensions relative to the ground-based or space-based navigation information.

[14] **Global positioning system (GPS)**: GPS is a satellite-based navigation system composed of a network of satellites placed into orbit by the United States Department of Defense (DOD). GPS was originally intended for military applications, but in the 1980s the government made the system available for civilian use. GPS works in all weather conditions, anywhere in the world, 24 hours a day. A GPS receiver must be locked onto the signal of at least four satellites to calculate a two-dimensional position (latitude and longitude) and track movement. With four or more satellites in view, the receiver can determine the user's three-dimensional position (latitude, longitude, and altitude). Other satellites must also be in view to offset signal loss and signal ambiguity.

NEW WORDS AND EXPRESSIONS

(1) balloon [bə'luːn] 气球
(2) airship ['eəʃɪp] 飞艇,飞船
(3) maneuverability [məˌnuːvərə'bɪlɪtɪ] 机动性,操纵性
(4) directionality [daɪrekʃə'nælətɪ] 方向性
(5) cockpit ['kɒkpɪt] 驾驶舱
(6) instrument ['ɪnstrʊmənt] 仪表
(7) gyro ['dʒaɪrəʊ] 陀螺,陀螺仪
(8) gimbal ['dʒɪmbəl] 万向支架,常平架
(9) orientation [ˌɔːrɪən'teɪʃn] 定位,定向
(10) navigation [ˌnævɪ'geɪʃn] 导航
(11) artificial [ˌɑːtɪ'fɪʃl] 人造的,人工的
(12) horizon [hə'raɪzn] 水平线,地平线;水平的
(13) compass ['kʌmpəs] 罗盘,指南针
(14) geomagnetic [ˌdʒiːəʊmæg'netɪk] 地磁的
(15) iron ['aɪən] 铁
(16) altimeter ['æltɪmiːtə(r)] 高度表

(17) jet [dʒet] 喷气式飞机;喷气式;喷射,喷出
(18) sound barrier 声障,音障
(19) reliability [rɪˌlaɪə'bɪlətɪ] 可靠性
(20) fuselage ['fuːzəlɑːʒ] 机身
(21) semi-monocoque 半硬壳式
(22) aluminum [ə'luːmɪnəm] 铝
(23) alloy ['ælɔɪ] 合金
(24) titanium [tɪ'teɪnɪəm] 钛
(25) cantilever ['kæntɪliːvə(r)] 悬臂,悬臂式梁
(26) sweptback ['swept͵bæk] 后掠(翼)
(27) cabin ['kæbɪn] 舱,座舱
(28) hijacking ['haɪdʒækɪŋ] 劫持(飞机)
(29) heading ['hedɪŋ] 航向
(30) AOA: angle of attack 迎角,攻角
(31) calibrate ['kælɪbreɪt] 校准
(32) increment ['ɪŋkrəmənt] 增量,增加
(33) VSI: vertical speed indicator 升降速度表
(34) GPS: global positioning system 全球定位系统
(35) VHF: very high frequency 甚高频
(36) VOR: omni direction radio range 甚高频全向信标台
(37) NDB: nondirectional beacon 无方向性信标
(38) ILS: instrument landing system 仪表着陆系统
(39) path [pɑːθ] 航路,轨迹;航迹
(40) satellite ['sætəlaɪt] 卫星,人造卫星
(41) orbit ['ɔːbɪt] 轨道
(42) DOD: United States Department of Defense 美国国防部
(43) military ['mɪlətrɪ] 军事的,军用的
(44) civilian [sə'vɪlɪən] 民用的
(45) offset ['ɒfset] 偏置,补偿
(46) ambiguity [ˌæmbɪ'gjuːətɪ] 模糊度,不定性

EXERCISE

1. Translate the followings into Chinese.

(1) **Balloon**. A lightweight aerostatic machine, usually spherical in shape and neither powered nor steerable. When the balloon is filled with hot air or with a gas that is lighter than air, it displaces more than it weighs, and the resulting aerostatic lift causes it to rise.

(2) **Airship**. An engine-driven lighter-than-air aircraft that can be steered.

(3) **Aircraft**. Any weight-carrying device designed to be supported by the air. Airplane, helicopters, gliders, and balloons are all types of aircraft.

(4) **Airplane**. An engine-driven, fixed-wing aircraft heavier than air, that is supported in flight by the dynamic reaction of the air against its wings.

(5) **Cockpit**. The portion of an aircraft or a spacecraft from which the flight crew controls the vehicle.

(6) **Fuselage**. The body, or central structural component of an airplane. The passengers and flight crew are housed in the fuselage, and the wings and tail are attached to it. In most single-engine airplanes, the engine and landing gear are attached to the fuselage.

(7) **Instrument**. A device using an internal mechanism to show visually or aurally the attitude, altitude, or operation of an aircraft or aircraft part. It includes electronic devices for automatically controlling an aircraft in flight.

(8) **Altimeter**. An aneroid barometer whose dial is calibrated in feet or meters above a reference pressure level. An altimeter measures the difference between the pressure of the air surrounding it and a reference pressure which is set on the barometric pressure dial on the face of the instrument. This pressure differences are expressed in feet or meters.

(9) **Sweptback wing**. An airplane wing in which the leading edge angles backward from the fuselage. The leading edge of the wing at the tip is farther back than the leading edge at the root.

(10) **Monocoque**. A single-shell type of aircraft structure in which all of the flight loads are carried in its outside skin. Modern aircraft have skins made of aluminum alloy, formed into compound curves, and riveted together into a structure that resembles an eggshell. There is a minimum of structure inside a monocoque skin.

2. Answer the following questions in English.

(1) What is the biggest difference between airships and balloons?

(2) What are the technical features of B737 series airplane?

(3) What are the technical features of A320 series airplane?

(4) Why does the cockpit door need to be secured?

(5) What are the aircraft performance instruments?

(6) What are the aircraft control instruments?

(7) What are the aircraft navigation instruments?

(8) How does the GPS receiver use satellite signals to determine the user's three-dimensional position?

SUPPLEMENTARY READING

[1] The B747-400 incorporates major aerodynamic improvements over earlier B747 models, including the addition of winglets to reduce drag, new avionics, a new flight deck and the latest in-flight entertainment systems. And, the B747 continues to be the world's fastest subsonic jetliner, cruising at Mach 0.855-or 85.5 percent of the speed of sound. With the lowest operating cost per ton-mile in the industry, the new-technology B747-400 Freighter is the all-cargo transport member of the B747-400 family. It can carry twice as much cargo, twice as far, as the competitor's leading freighter.

[2] The B777-300ER extends the B777 family's span of capabilities, bringing twin-engine efficiency and reliability to the long-range market. The airplane carries 365 passengers up to 7,930 nautical miles (14,685 km). Boeing incorporated several performance enhancements for the B777-300ER, extending its range and payload capabilities. Excellent performance during flight testing, combined with engine efficiency improvements and design changes that reduce drag and airplane weight, contributed to the increased capability. The benefits were also applied during development of the B777-200LR and the B777 freighter.

[3] In conjunction with an announcement that the B787 will be made primarily of composite materials, Boeing released a new image of the super efficient airplane. Work on development of the airplane continues on schedule. More than 40 airlines from around the world are engaged in dialogues with Boeing about the B787.

[4] The A330 is the most cost efficient wide-body twin aircraft in operation. The A330 Family has five members—the A330-200, A330-300, A330-200F, A330 Prestige and A330 MRTT—which cover all market segments with one twin-engine aircraft type. The combination of low operating costs, high efficiency, flexibility and optimised performance makes the A330 family popular with an ever-increasing operator base. Continuous improvement brings the latest technology onboard the A330. Commonality is further shared across the Airbus product line of widebody and single-aisle aircraft through the use of fly-by-wire controls and cockpits with similar layouts. The A330 provides exceptional operational flexibility at lowest cost, with right-sized cabins for passenger, freight, VIP and multi-role military applications. It combines Airbus' flight-by-wire technology and modern onboard systems with high aerodynamic efficiency for a highly productive aircraft.

[5] The A350 XWB jetliner is shaping the future of air travel by offering a complete family of new-generation aircraft that is best suited to the market's requirements for size, range, revenue generation, passenger comfort and the environment. The A350 family provides true long-range capability with seating capacities from 250 to 400-plus passengers. This enables airlines to best match their A350 XWB fleets to route capacity

demands, guaranteeing optimum revenue potential and excellent operating efficiency. Over 70 percent of the A350 XWB's weight-efficient airframe is made from advanced materials, combining 53 percent of composite structures with titanium and advanced aluminum alloys. The aircraft's innovative all-new carbon fiber reinforced plastic (CFRP) fuselage results in lower fuel consumption, as well as easier maintenance. The aircraft family concept, proven by Airbus with its other jetliners, also ensures optimal efficiency through the A350 XWB's commonality in engines, systems and spare parts, while also enabling pilots to fly all three versions with a single type rating.

[6] The double-deck A380 is the world's largest commercial aircraft flying today, with capacity to carry 525 passengers in a comfortable three-class configuration, and up to 853 in a single-class configuration that provides wider seats than its competitor. Overall, the A380's two decks offer 50% more floor surface than any other high-capacity aircraft. With its range of 8,300 nautical miles, the A380 is the ideal solution to alleviate traffic congestion at busy airports, while coping with market growth. It has two full-length passenger levels with true widebody dimensions: a main deck and an upper deck, which are linked by fixed stairs forward and aft.

TOPIC 1：科技英语特点

科技英语是指在自然科学和工程技术方面的科学著作、论文、教科书、科技报告和学术讲演中所使用的英语。随着科技的发展与全球经济一体化的逐步深入,科技英语越来越彰显出其重要性,因此世界上许多国家都设立了科技英语研究机构,并在大学中设立了相关专业。科学家钱三强曾指出:"科技英语在许多国家已经成为现代英语的一个专门的新领域。"

科技英语在大学生课程学习阶段通常也称为"专业英语"。大学生在经过基础英语学习之后,基本上已经掌握了英语的常用语法,并拥有4000以上的词汇量,具备了较扎实的英语基础。进入大学三年级之后,随着专业课的进一步学习,学生的专业知识技能也开始逐步加强。具备以上两个条件后,学生应进行专业英语的训练,在保证30万词以上阅读量的基础上,对本专业英文资料的阅读才算达到基本的要求。换言之,掌握专业英语技能是大学基础英语学习的主要目的之一,是一种素质上的提高,直接关系到学生的求职和毕业后的工作能力,而这一点在民航业显得尤为重要。

科技英语有以下特点:

(1) 科技英语不像普通英语那样具有感性形象思维,不具感情色彩,其目的是使读者容易理解而不产生太多的想象,通常不用比喻、排比、夸张等修辞手法,而是要准确表达客观规律,按逻辑思维清晰地描述问题。

(2) 科技英语的词汇意义比较专一、稳定,特别是大量的专业名词,其词义固定、专一,即使是像 do,take,make 这样的多意义普通动词,在科技英语中它们的词义也比较固定,其表达方式也比较容易理解。

(3) 科技英语主要是一种书面语言,它要求严谨、简洁,不要求在文中堆积华丽的辞藻,也不要求考虑朗读和吟诵。

(4) 科技英语词汇具有国际性,据有关统计,70%以上的科技英语词汇来自拉丁语、希腊语。

(5) 经常使用正式规范的书面动词来替代具有同样意义的口语化的动词或动词短语。

(6) 在语法结构上,科技英语大量使用被动语态。科技英语使用被动语态可以使描述减少主观色彩,增强客观性,并通过隐去人称主语而使句子尽可能简洁。

(7) 大量使用名词或名词短语。

(8) 大量使用非谓语动词短语及分词短语。

(9) 用 It 做形式主语替代后面 that 所引导的作为主句真正主语的从句。

(10) 常用 It 做形式主语替代句子后面作真正主语的动词不定式短语。

(11) 多用介词词组来表示用什么方法、数据、资料、材料、标准等。

(12) 常包含两个及两个以上从句的长句。

1. 词汇特点

科技词汇的词形一般较长,多源于希腊语和拉丁语。据美国科技英语专家 Oscar E.

Nybaken 统计,在一万个普通英语的词汇中,约有 46% 的词汇源于拉丁语,7.2% 源于希腊语。尤其在专业性极强的科技英语词汇中,这种比例就更高,例如:dynamics,electric,physics,pneumatic 等希腊语和拉丁语。它们之所以能成为科技词汇的基本来源,是因为这两种语言都是"死"语言,不会由于社会的发展而引起词义的变化,也不会因词的多义引起歧义。

词汇可以分为专业词汇、次专业词汇和非专业词汇。

(1) 专业词汇:某个专业特有的词汇,专业性强,词义狭窄和单一,如:
compressor(压缩机),turbine(涡轮),refrigeration(制冷),airfoil(翼型)。

(2) 次专业词汇:很多专业和学科共有词汇,在不同专业和学科中代表不同词义,如:
power:幂,乘方;动力,功率;效率;电动的,用电力发动的
powerplant:发电厂,发电站;动力装置,发电机
oil:石油,原油;滑油;食用油
condenser:冷凝器;电容器

(3) 非专业词汇:在普通英语或非专业英语中使用较少,但实际却属于非专业英语的词汇,如:
take in→absorb,find out→discover,turn round→rotate,change→convert。

除此之外,专业词汇采用大量的合成词、派生词和缩略词。

2. 语法特点

科技英语在表达时力求做到客观性、准确性以及精练性,这就决定了它的语法具有以下特点。

1) 广泛使用被动语态(客观性)

> The emergency equipment **is installed** in the aircraft for the safety of the passengers and the crew。
> 飞机上装有应急设备用于保证乘客和机组的安全。

2) 广泛使用非谓语形式(精练性)

非谓语动词在句子中可以起到名词、形容词或者副词的作用,动词的非谓语形式分为动名词、分词、动词不定式。

(1) 动名词:用动名词短语取代时间从句或简化时间陈述句。

> We must do various experiments before a new electronic product is designed.
> 在设计一个新的电子产品之前,我们必须做各种实验。
> Before **designing** a new electronic product, we must do various experiments.

动名词还经常用作主语:

> **Cleaning** methods should be closely observed or replaced by procedure specified by the manufacturer of equipment substitutes.
> 清洁应当严格遵守手册中的方法,或者是生产厂家规定的等效替代程序。

(2) 分词。

过去分词短语替代从句中的被动语态:

> The kinetic energy is the energy which is caused by the velocity of the molecules.
> 动能是由于分子运动产生的能量。
> The kinetic energy is the energy **caused** by the velocity of the molecules.

现在分词短语替代从句中的主动语态：

> The transistor, which is working with correctly polarities, can work as an amplifier.
> 在正确电源极性下工作的晶体管作用就像放大器。
> The transistor **working** with correctly polarities can work as an amplifier.

（3）不定式：用不定式短语来替代表示目的和功能的从句或语句。

> **To assure** proper system operation and **to avoid** damage to none-metallic components of the hydraulic system, the correct fluid must be used.
> 为确保液压系统正常工作以及防止非金属元件损坏，必须采用正确牌号的液压油。

3）使用省略句

省略成分：状语从句中的主语、全部或部分谓语、定语从句中的关系代词 which 和 that、从句中的助词等，还常用介词短语替代从句。

> Defects in the component, if any, can be detected by ultrasonic waves.
> 元件中的缺陷，如果有的话，可用超声波检测出来。

其他常用的省略形式有：

as already discussed 前已讨论　　as explained before 前已解释
as described above 如上所述　　　if possible 如果可能
if so 倘若如此　　　　　　　　　as previously mentioned 前已提到
when needed 必要时　　　　　　where feasible 在切实可行的场合
where possible 在可能的情况下

4）频繁使用 it 句型和祈使句（准确、精练）

（1）it 句型：it 充当形式主语，避免句子"头重脚轻"。常见形式有：

It is very important (possible, necessary, neutral) to…
It takes very much time learning…
It is clear (possible, necessary, neutral) to…
It happened that…
It must be admitted that…

如：It is not necessary for air to pass through an enclosed tube for Bernoulli principle to apply.

（2）祈使句：没有主语，精练。

Make sure that the ground safety locks are in position on the landing gear.
Put the recirculation fan lugs in position and install the washers, the bolts and the nuts.

5)使用复杂长句

为了完整、准确地表达事物内在联系,使用大量从句。

Filters may be located within the reservoir in the pressure line, in the return line or in any other location **where** the designer of the system decides **that** they are needed to safeguard the hydraulic system against impurities.

6)多采用后置形容词短语做定语

形容词短语代替定语从句做后置定语,使句子简洁紧凑,不累赘。

The instruments present include some digital ones **relative to** NDT.

UNIT 2

PRINCIPLE OF FLIGHT

[1] Lift, weight, thrust, and drag are forces that act upon all aircraft in flight. It's illustrated in Fig. 2-1. Understanding how these forces work and knowing how to control them with the use of power and flight controls are essential to flight. Lift opposes the downward force of weight, is produced by the dynamic effect of the air acting on the airfoil, and acts perpendicular to the flight path through the center of lift. Weight is the combined load of the aircraft itself, the crew, the fuel, and the cargo or baggage. Weight pulls the aircraft downward because of the force of gravity. It opposes lift, and acts vertically downward through the aircraft's center of gravity (CG). Thrust is the forward force produced by the powerplant/propeller or rotor. It opposes or overcomes the force of drag. Drag is a rearward, retarding force caused by disruption of airflow by the wing, rotor, fuselage, and other protruding objects. Drag opposes thrust, and acts parallel to the relative wind.

[2] **Lift**: It is not necessary for air to pass through an enclosed tube for Bernoulli's Principle to apply. Any surface that alters airflow causes a venturi effect. For example, if the upper portion of the tube is removed, the venturi effect applies to air flowing along the lower section of the tube. Velocity above the curvature is increased and pressure is decreased. You can see how the venturi effect works on a wing, or airfoil, if you picture an airfoil inset in the curved part of the tube. It's illustrated in Fig. 2-2.

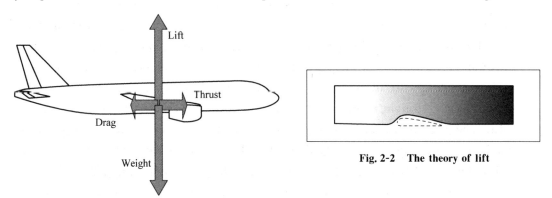

Fig. 2-2 The theory of lift

Fig. 2-1 The four aerodynamic forces

[3] Air flowing over the top of an airfoil reaches the trailing edge in the same amount of time as air flowing along the relatively flat bottom. Since both the upper and lower surfaces pass through a block of air at the same speed, the air flowing over the curved upper surface travels farther. This means it must go faster, resulting in lower pressure above the airfoil and a greater pressure below. An airfoil is specially designed to produce a reaction with the air that passes over it. This difference in pressure is the main source of lift.

[4] **Weight**: Gravity is the pulling force that tends to draw all bodies to the center of the earth. The center of gravity (CG) may be considered as a point at which all the weight of the airplane is concentrated. If the airplane is supported at its exact center of gravity, it would balance in any attitude. It will be noted that center of gravity is of major importance in an airplane, for its position has a great bearing upon stability.

[5] The location of the center of gravity is determined by the general design of each particular airplane. The designers determine how far the center of pressure (CP) will travel. They then fix the center of gravity forward of the center of pressure for the corresponding flight speed in order to provide an adequate restoring moment to retain flight equilibrium.

[6] Weight has a definite relationship with lift, and thrust with drag. This relationship is simple, but important in understanding the aerodynamics of flying. Lift is the upward force on the wing acting perpendicular to the relative wind. Lift is required to counteract the airplane's weight. This weight (gravity) force acts downward through the airplane's center of gravity. In stabilized level flight, when the lift force is equal to the weight force, the airplane is in a state of equilibrium and neither gains nor loses altitude.

[7] **Thrust**: Before the airplane begins to move, thrust must be exerted. It continues to move and gain speed until thrust and drag are equal. In order to maintain a constant airspeed, thrust and drag must remain equal, just as lift and weight must be equal to maintain a constant altitude. If in level flight, the engine power is reduced, the thrust is lessened, and the airplane slows down. As long as the thrust is less than the drag, the airplane continues to decelerate until its airspeed is insufficient to support it in the air.

[8] Likewise, if the engine power is increased, thrust becomes greater than drag and the airspeed increases. As long as the thrust continues to be greater than the drag, the airplane continues to accelerate. When drag equals thrust, the airplane flies at a constant airspeed. Straight and level flight may be sustained at speeds from very slow to very fast. The pilot must coordinate angle of attack and thrust in all speed regimes if the airplane is to be held in level flight. Roughly, these regimes can be grouped in three categories: low-speed flight, cruising flight, and high-speed flight (Fig.2-3).

[9] When the airspeed is low, the angle of attack must be relatively high to increase lift if the balance between lift and weight is to be maintained. If thrust decreases and airspeed decreases, lift becomes less than weight and the airplane will start to descend.

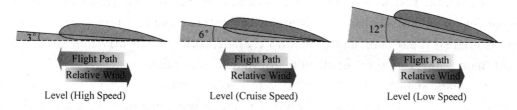

Fig. 2-3 Angle of attack at various speeds

To maintain level flight, the pilot can increase the angle of attack an amount which will generate a lift force again equal to the weight of the airplane and while the airplane will be flying more slowly, it will still maintain level flight if the pilot has properly coordinated thrust and angle of attack.

[10] **Drag**: Drag is the force that resists movement of an aircraft through the air. There are two basic types: parasite drag and induced drag. The first one is called parasite because it in no way functions to aid flight, while the second one, induced drag, is a result of an airfoil developing lift.

[11] Parasite drag is composed of all the forces that work to slow an aircraft's movement. As the term parasite implies, it is the drag that is not associated with the production of lift. This includes the displacement of the air by the aircraft, turbulence generated in the airstream, or a hindrance of air moving over the surface of the aircraft and airfoil. There are three types of parasite drag: form drag, interference drag, and skin friction.

- Form drag is created by any structure which protrudes into the relative wind. The amount of drag created is related to both the size and shape of the structure. For example, a square strut creates substantially more drag than a smooth or rounded strut. Streamlining reduces form drag.
- Skin friction drag is caused by the roughness of the airplane's surfaces. Even though these surfaces may appear smooth, under a microscope they may be quite rough. A thin layer of air clings to these rough surfaces and creates small eddies which contribute to drag.
- Interference drag occurs when varied currents of air over an airplane meet and interact. This interaction creates additional drag. One example of this type of drag is the mixing of the air where the wing and fuselage join.

[12] Induced drag is the main by-product of the production of lift. It is directly related to the angle of attack of the wing. The greater the angle, the greater the induced drag. Since the wing usually is at a low angle of attack at high speed, and a high angle of attack at low speed, the relationship of induced drag to speed also can be plotted. Over the past several years the winglet has been developed and used to reduce induced drag.

[13] An aircraft must have sufficient stability to maintain a uniform flight path and

recover from the various upsetting forces. Stability is the inherent quality of an aircraft to correct for conditions that may disturb its equilibrium, and to return to or to continue on the original flight path, and to tend to cause it to fly (hands off) in a straight and level flight path. Two types of stability are static and dynamic.

[14] **Static stability**: Static stability refers to the initial tendency, or direction of movement, back to equilibrium. The three types of static stability are defined by the character of movement following some disturbance from equilibrium. Positive static stability exists when the disturbed object tends to return to equilibrium. Negative static stability or static instability exists when the disturbed object tends to continue in the direction of disturbance. Neutral static stability exists when the disturbed object has neither the tendency to return nor continue in the displacement direction, but remains in equilibrium in the direction of disturbance. These three types of stability are illustrated in Fig. 2-4.

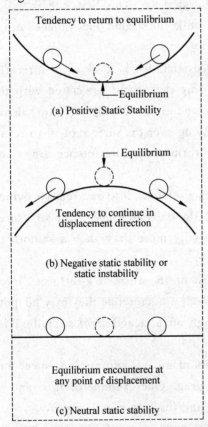

Fig. 2-4　Static stability

[15] While static stability deals with the tendency of a displaced body to return to equilibrium, dynamic stability refers to the aircraft response to time. If an object is disturbed from equilibrium, the time history of the resulting motion defines the dynamic stability of the object. In general, an object demonstrates positive dynamic stability if the amplitude of motion decreases with time. If the amplitude of motion increases with time, the object is said to possess dynamic instability.

[16] Any aircraft must demonstrate the required degrees of static and dynamic stability. If an aircraft were designed with static instability and a rapid rate of dynamic instability, the aircraft would be very difficult, if not impossible, to fly. Usually, positive dynamic stability is required in an aircraft design to prevent objectionable continued oscillations of the aircraft.

[17] **Longitudinal stability**: When an aircraft has a tendency to keep a constant angle of attack with reference to the relative wind-that is, when it does not tend to put its nose down and dive or lift its nose and stall-it is said to have longitudinal stability. Longitudinal stability refers to motion in pitch. The horizontal stabilizer is the primary surface which controls longitudinal stability. The action of the stabilizer depends upon the speed and angle of attack of the aircraft.

[18] Fig. 2-5 illustrates the contribution of tail lift to stability. If the aircraft changes its angle of attack, a change in lift takes place at the aerodynamic center (center of pressure) of the horizontal stabilizer. Under certain conditions of speed, load, and angle of attack, the flow of air over the horizontal stabilizer creates a force which pushes the tail up or down. When conditions are such that the airflow creates equal forces up and down, the forces are said to be in equilibrium. This condition is usually found in level flight in calm air.

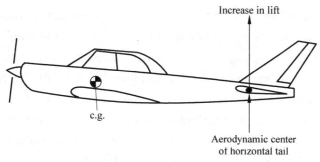

Fig. 2-5 Producing tail lift

[19] **Directional stability**: Stability about the vertical axis is referred to as directional stability. The aircraft should be designed so that when it is in straight and level flight it remains on its course heading even though the pilot takes his hands and feet off the controls. If an aircraft recovers automatically from a skid, it has been well designed and possesses good directional balance. The vertical stabilizer is the primary surface which controls directional stability.

[20] As shown in Fig. 2-6, when an aircraft is in a sideslip or yawing, the vertical tail experiences a change in angle of attack with a resulting change in lift (not to be confused with the lift created by the wing). The change in lift, or side force, on the vertical tail creates a yawing moment about the center of gravity which tends to return the aircraft to its original flight path.

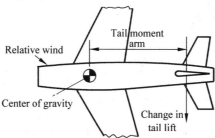

Fig. 2-6 Contribution of vertical tail to directional stability

[21] Sweptback wings aid in directional stability. If the aircraft yaws from its direction of flight, the wing which is farther ahead offers more drag than the wing which is aft. The effect of this drag is to hold back the wing which is farther ahead, and

to let the other wing catch up. Directional stability is also aided by using a large dorsal fin and a long fuselage. The high Mach number of supersonic flight reduces the contribution of the vertical tail to directional stability. To produce the required directional stability at high Mach numbers, a very large vertical tail area may be necessary. Ventral (belly) fins may be added as an additional contribution to directional stability.

[22] **Lateral stability**: We have seen that pitching is motion about the aircraft's lateral axis and yawing is motion about its vertical axis. Motion about its longitudinal (fore and aft) axis is a lateral or rolling motion. The tendency to return to the original attitude from such motion is called lateral stability. The lateral stability of an airplane involves consideration of rolling moments due to sideslip. A sideslip tends to produce both a rolling and a yawing motion. If an airplane has a favorable rolling moment, a sideslip will tend to return the airplane to a level flight attitude.

[23] The principal surface contributing to the lateral stability of an airplane is the wing. The effect of the geometric dihedral of a wing is a powerful contribution to lateral stability. As shown in Fig. 2-7, a wing with dihedral develops stable rolling moments with sideslip. With the relative wind from the side, the wing into the wind is subject to an increase in angle of attack and develops an increase in lift. The wing away from the wind is subject to a decrease in angle of attack and develops less lift. The changes in lift effect a rolling moment tending to raise the windward wing.

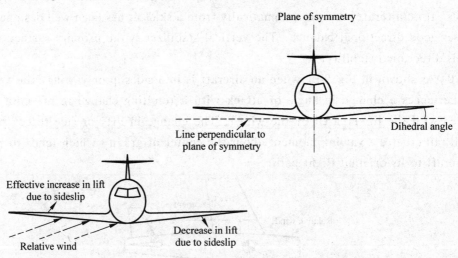

Fig. 2-7 Contribution of dihedral to lateral stability

[24] When a wing is swept back, the effective dihedral increases rapidly with a change in the lift coefficient of the wing. Sweptback is the angle between a line perpendicular to the fuselage center line and the quarter chord of each wing airfoil section. Sweepback in combination with dihedral causes the dihedral effect to be excessive. As shown in Fig. 2-8, the swept wing aircraft in a sideslip has the wing that is into the wind operating with an effective decrease in sweepback, while the wing out of

the wind is operating with an effective increase in sweepback.

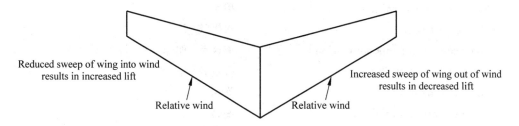

Fig. 2-8 Effect of sweepback on lateral stability

[25] The wing into the wind develops more lift, and the wing out of the wind develops less. This tends to restore the aircraft to a level flight attitude. The amount of effective dihedral necessary to produce satisfactory flying qualities varies greatly with the type and purpose of the aircraft. Generally, the effective dihedral is kept low, since high roll due to sideslip can create problems. Excessive dihedral effect can lead to Dutch Roll, difficult rudder coordination in rolling maneuvers, or place extreme demands for lateral control power during crosswind takeoff and landing.

NEW WORDS AND EXPRESSIONS

(1) thrust[θrʌst]　　　　　　　　推力,拉力
(2) gravity['grævɪtɪ]　　　　　　　重力
(3) curvature['kɜːvəˌtʃuə]　　　　弯曲,弯度,曲率
(4) airfoil['eəfɔɪl]　　　　　　　　翼型,机翼
(5) cruising[kruːzɪŋ]　　　　　　 巡航的
(6) propel[prə'pel]　　　　　　　推进,推动
(7) speed regime　　　　　　　　速度包线
(8) aerodynamics[ˌeərəudaɪ'næmɪks]　空气动力学,空气动力特性
(9) mass[mæs]　　　　　　　　　质量
(10) altitude['æltɪtjuːd]　　　　　 高度
(11) equilibrium[ˌiːkwə'lɪbriːəm]　　平衡
(12) descend[dɪ'send]　　　　　　下降
(13) cambered['kæmbəd]　　　　 曲面的,弧形的
(14) parasite drag　　　　　　　　寄生阻力,废阻力
(15) induced[ɪn'djuːsd]　　　　　诱导的
(16) disruption[dɪs'rʌpʃən]　　　　中断
(17) form[fɔːm]　　　　　　　　 形状
(18) friction['frɪkʃən]　　　　　　 摩擦
(19) interference[ˌɪntə'fɪərəns]　　　干扰

(20) streamline ['stri:mlaɪn]　　流线型,流线；使成流线型
(21) roughness ['rʌf.nɪs]　　粗糙
(22) level flight　　水平飞行
(23) acceleration [æk,selə'reɪʃən]　　加速度,加速
(24) oscillator ['ɒsɪleɪtə]　　振荡器,振子
(25) gust [gʌst]　　阵风
(26) deflection [dɪ'flekʃən]　　偏转
(27) static stability　　静稳定性
(28) disturbance [dɪs'tə:bəns]　　干扰,打扰
(29) negative ['negətɪv]　　负的
(30) static instability　　静态不稳定性
(31) neutral ['nju:trəl]　　中立的
(32) vertical axis　　垂直轴,立轴
(33) tendency ['tendənsɪ]　　倾向
(34) displacement [dɪs'pleɪsmənt]　　位移,排量
(35) dynamic stability　　动稳定性
(36) amplitude ['æmplɪ,tju:d]　　振幅,波幅
(37) rate [reɪt]　　比率,等级
(38) reference ['refrəns]　　参考,涉及
(39) stall [stɔ:l]　　失速,气流分离
(40) longitudinal stability　　纵向稳定性
(41) horizontal stabilizer　　水平安定面
(42) aerodynamic center　　气动中心
(43) calm [kɑ:m]　　平静的,沉着的
(44) directional [dɪ'rekʃənəl]　　方向的
(45) dorsal fin　　背鳍
(46) ventral (belly) fins　　腹鳍
(47) attitude ['ætɪtju:d]　　姿态
(48) dihedral [daɪ'hi:drəl]　　上反的
(49) perpendicular [,pə:pən'dɪkjələ]　　垂直的
(50) crosswind ['krɔswɪnd]　　侧风
(51) sweptback ['swept,bæk]　　后掠的,向后倾斜的

EXERCISE

1. Translate the followings into Chinese.

(1) **Drag.** An aerodynamic force acting in the same plane as the relative wind striking an airfoil. Two basic types of drag act on an aircraft in flight: induced

drag and parasite drag.

(2) **Gravity.** The force of attraction between objects. The force of gravity is directly proportional to the mass of the objects (the greater the mass, the stronger the pull of gravity), and inversely proportional to the square of the distance between the objects (the farther apart the objects, the weaker the pull of gravity).

(3) **Lift.** An aerodynamic force caused by air flowing over a specially shaped surface called an airfoil. The airfoil is curved in such a way that the air flowing over the upper surface finds the surface falling away from it. The air is being pressed onto the surface by the air above it, and in order for it to remain on the surface, it must speed up. As the air speeds up, its pressure drops, and the air above it flows down to fill the low-pressure area.

(4) **Thrust.** The forward aerodynamic force produced by a propeller, fan, or turbojet engine as it forces a mass of air to the rear, behind the airplane. A propeller produces its thrust by accelerating a large mass of air by a relatively small amount. A turbojet engine produces its thrust by accelerating a smaller mass of air by a much larger amount. The mass and acceleration of the air moved by a fan is between those of the propeller and the jet.

(5) **Stability.** The ability of an object, such as a ship or an aircraft, to maintain equilibrium or resume its original position after displacement, as by the sea or strong winds.

(6) **Angle of attack.** The angle of the wing chord in relation to the free stream airflow as it moves forward through the air.

(7) **Stall.** In fluid dynamics, a stall is a reduction in the lift coefficient generated by a foil as angle of attack increases. This occurs when the critical angle of attack of the foil is exceeded. The critical angle of attack is typically about 15 degrees, but it may vary significantly depending on the fluid, foil, and Reynolds number.

(8) **Dihedral.** The upward or downward inclination of an aircraft wing from true horizontal.

(9) **Chord.** A straight line connecting the leading and trailing edges of an airfoil.

(10) **Sideslip.** To fly sideways and downward in an airplane along the lateral axis to reduce altitude without gaining speed or as the result of banking too deeply.

2. Answer the following questions in English.
(1) What are the four forces acting on an aircraft?
(2) How is the lift generated?
(3) What are the three kinds of parasite drag acting on the sunsonic aircraft?
(4) What are the three types of static stability?

(5) What are the three types of dynamic stability of an aircraft? Which is the primary surface contributing to each type of dynamic stability respectively?

SUPPLEMENTARY READING

[1] Stability is the inherent quality of an aircraft to correct for conditions that may disturb its equilibrium, and to return to or to continue on the original flightpath. It is primarily an aircraft design characteristic. The flightpaths and attitudes an aircraft flies are limited by the aerodynamic characteristics of the aircraft, its propulsion system, and its structural strength. These limitations indicate the maximum performance and maneuverability of the aircraft. If the aircraft is to provide maximum utility, it must be safely controllable to the full extent of these limits without exceeding the pilot's strength or requiring exceptional flying ability. If an aircraft is to fly straight and steady along any arbitrary flightpath, the forces acting on it must be in static equilibrium. The reaction of any body when its equilibrium is disturbed is referred to as stability. The two types of stability are static and dynamic.

[2] Static stability refers to the initial tendency, or direction of movement, back to equilibrium. In aviation, it refers to the aircraft's initial response when disturbed from a given AOA, slip, or bank.

[3] Positive static stability: the initial tendency of the aircraft to return to the original state of equilibrium after being disturbed.

[4] Neutral static stability: the initial tendency of the aircraft to remain in a new condition after its equilibrium has been disturbed.

[5] Static stability: the initial tendency of the aircraft to continue away from the original state of equilibrium after being disturbed.

[6] Static stability has been defined as the initial tendency to return to equilibrium that the aircraft displays after being disturbed from its trimmed condition. Occasionally, the initial tendency is different or opposite from the overall tendency, so a distinction must be made between the two.

[7] Dynamic stability refers to the aircraft response over time when disturbed from a given AOA, slip, or bank. This type of stability also has three subtypes:

- Positive dynamic stability: over time, the motion of the displaced object decreases in amplitude and, because it is positive, the object displaced returns toward the equilibrium state.
- Neutral dynamic stability: once displaced, the displaced object neither decreases nor increases in amplitude. A worn automobile shock absorber exhibits this tendency.
- Negative dynamic stability: over time, the motion of the displaced object

increases and becomes more divergent.

[8] Stability in an aircraft affects two areas significantly:
- Maneuverability: the quality of an aircraft that permits it to be maneuvered easily and to withstand the stress imposed by maneuvers. It is governed by the aircraft's weight, inertia, size and location of flight controls, structural strength, and powerplant. It also is an aircraft design characteristic.
- Controllability: the capability of an aircraft to respond to the pilot's control, especially with regard to flightpath and attitude. It is the quality of the aircraft's response to the pilot's control application when maneuvering the aircraft, regardless of its stability characteristics.

[9] Spiral instability exists when the static directional stability of the aircraft is very strong as compared to the effect of its dihedral in maintaining lateral equilibrium. When the lateral equilibrium of the aircraft is disturbed by a gust of air and a sideslip is introduced, the strong directional stability tends to yaw the nose into the resultant relative wind while the comparatively weak dihedral lags in restoring the lateral balance. Due to this yaw, the wing on the outside of the turning moment travels forward faster than the inside wing and, as a consequence, its lift becomes greater. This produces an overbanking tendency which, if not corrected by the pilot, results in the bank angle becoming steeper and steeper. At the same time, the strong directional stability that yaws the aircraft into the relative wind is actually forcing the nose to a lower pitch attitude. A slow downward spiral begins which, if not counteracted by the pilot, gradually increases into a steep spiral dive. Usually the rate of divergence in the spiral motion is so gradual that the pilot can control the tendency without any difficulty.

[10] All aircraft are affected to some degree by this characteristic, although they may be inherently stable in all other normal parameters. This tendency explains why an aircraft cannot be flown "hands off" indefinitely.

[11] Much research has gone into the development of control devices (wing leveler) to correct or eliminate this instability. The pilot must be careful in application of recovery controls during advanced stages of this spiral condition or excessive loads may be imposed on the structure. Improper recovery from spiral instability leading to inflight structural failures has probably contributed to more fatalities in general aviation aircraft than any other factor. Since the airspeed in the spiral condition builds up rapidly, the application of back elevator force to reduce this speed and to pull the nose up only "tightens the turn", increasing the load factor. The results of the prolonged uncontrolled spiral are inflight structural failure or crashing into the ground, or both. The most common recorded causes for pilots who get into this situation are: loss of horizon reference, inability to control the aircraft by reference to instruments, or a combination of both.

TOPIC 2：专业英语翻译标准与技巧

1. 科技英语翻译标准

翻译标准是评价译文质量的尺度,综合起来有以下3点:

(1) 翻译首先要做到忠实、准确,要"信"。也就是说译文应避免漏译或者错译,要忠实、准确地传达原文的内容,既不能歪曲,也不能任意增减。同时,在表达上保持原作的风格和文体。

> The importance of safety in civil aviation can not be overestimated.
> 初译:安全在民用航空中的重要性不能被估计过高。
> 更正:安全在民用航空中的重要性怎么高估都不过分。

(2) 要做到通顺、流畅,要"达"。这是指译文应通顺易懂,符合汉语的表达规范,要和原作同样的流利自如。同时,在翻译中要避免死译和硬译,以便他人理解。

> This possibility was supported to a limited extend in the wind-tunnel test.
> 初译:这一可能性在风洞实验中在有限的程度上被支持了。
> 更正:一定程度上这一可能性在风洞试验中得到了证实。

> The difference between air cycle cooling system and vapor cycle system do not stop there.
> 初译:空气循环制冷系统和蒸发循环制冷系统的差别不停留在那里。
> 更正:空气循环制冷系统和蒸发循环制冷系统的差别不止如此。

忠实和通顺是辩证统一的关系,要防止对"忠实"的片面理解,一味追求形式上的相似,会造成逐字翻译,产生翻译上的形式主义。

> Protect reworked area against corrosion and dissimilar metal per SRM chapter 51.
> 初译:根据结构修理手册51章保护再加工区域防止腐蚀和不同金属。
> 更正:按照结构修理手册51章保护修理区域,以防止腐蚀和不同金属相接触。

另外,也要防止片面理解"通顺"的要求,过分强调译文的流畅而不受原文的约束,添枝加叶,造成翻译上的自由主义。

> This technique provides a solution with the longest range and the maximum data rate for user.
> 初译:这一技术提供给用户具有目前最为广泛的传输范围和最大的数据传输速度的最为有效的解决方案。
> 更正:这一技术为用户提供了具有最大数据传输距离和速度的方案。

(3) 译文在忠实和通顺的基础上还应注意文采,要"雅"。要讲究修辞,使译文在逻辑上严谨且流畅,语言上优美且易懂,否则就会"言之无文,行之不远"而失去读者。如:在译文中适当的添加一些承上启下的连接词,灵活使用汉语中的成语或俗语,如 vice versa 译为

"反之亦然",it is well known that...翻译为"众所周知……"等。

此外,无论翻译和阅读,都希望有较高的翻译速度,速度越快,对信息的获取量就越大,效率也就越高。

2. 科技英语翻译的方法

要想保证科技英语长句翻译的质量,不但要有扎实的专业知识和中英文能力,还要对科技英语长句的句式特点和翻译原则有所了解。此外,还需要掌握常用的科技英语长句的翻译方法,在翻译科技英语长句时有法可依、事半功倍。科技英语长句的翻译可以用以下方法:

1) 顺译法

顺译法就是在判断原文的词序及逻辑结构与汉语一致时,可以采用基本一致的顺序进行翻译。在翻译科技英语长句时,可以发现并非每一个英语长句的结构都有别于汉语结构。当遇上语法结构、词序及逻辑关系与汉语非常接近的句子时,基本上可按原文的词序和逻辑结构依次译出。在使用顺译法的时候,要先把句子拆分成意义单位,然后依次译出。但是,顺译法并不是每一个词都按照原句顺序。汉语与英语属于不同语系,两者的构句有差别,需要变化的则加以改变,使之更符合汉语的表达习惯。

> Once an engine has been selected, the propulsion engineering tasks are to design the air inlet for the engine, and to assure the satisfactory physical and aerodynamic integration of the inlet, engine, and exhaust nozzle or the engine nacelles with the rest of the airframe.
>
> 一旦选定某型发动机,推进系统工程的任务就是为该发动机设计进气道,并确保进气道、发动机本体、排气喷口或发动机短舱能够与机体的其余部分达到良好的物理和气动融合。

2) 逆译法

虽然顺译法可以用于那些在词序及逻辑关系与汉语十分相似的句子上,但是科技英语长句的时间顺序或逻辑顺序往往与汉语差别较大,有时甚至正好相反。此时,需要改变原文语序,甚至逆着原文顺序翻译,使译文更加符合汉语表达的方式。在翻译实践中,可以经常发现一些科技英语长句的表达方式不符合汉语习惯的情形,需要较多地使用逆译法。在翻译科技英语长句的时候,可以发现逆译法运用最多的就是状语从句顺序的调整。英语结构灵活,一些状语成分位置多变,可前可后,而汉语遵守因果和时间空间顺序,基本上状语都是前置的,这也是由汉语的铺垫在前、中心在后的叙述规律所决定的。在翻译时,需要注意英汉两种语言的这一差异,并进行必要的句序调整,这样才能使译文更加通顺流畅,有助于原文信息准确有效地传递。

> Use only the ground service network to supply electricity because the avionics ventilation system is not serviceable.
>
> 由于电子通风系统不可用,所以只能使用地面勤务网供电。

3) 分译法

分译法是翻译科技英语长句最常用的翻译方法。由于汉语多使用较短的句式结构,翻译时可以将科技英语长句分开来翻译。当短语或从句与句子中的其他成分的联系并不十分紧密时,或者当科技英语长句的句式结构关系过于复杂,翻译成一句汉语长句会显得冗长拖沓且语义混乱时,可以使用分译法,即采取化整为零的方法,使译文更符合中文的表达习惯。翻译时可以将原句在连接词处断开,或按照意群切分,译成几个独立的汉语短句。

> When the hydraulic pressure is relieved, the retracting springs in the automatic adjusters force the pistons to retract into the chambers and the hydraulic fluid is bled through the brake hose connection to return line.
> 当液压压力释放时,自动调节器内的复位弹簧迫使活塞收回到作动筒腔,液压油通过刹车软管连接件回到回油管路。

3. 被动句的翻译

科技文章侧重描述和推理,强调客观准确,所以谓语大量地采用被动语态,以避免过多使用第一、二人称而引起主观臆断的印象。英语和汉语都有被动语态,但这两种语言在运用和表达上却不尽相同,因此翻译时必须对被动句作适当的灵活处理。

(1) 英语中某些着重被动动作的被动句为突出其被动意义,可直接译为汉语的被动句。翻译时,最常见的方式是在谓语动词前加"被"。但过多地使用同一个词会使译文缺乏文采,因而应根据汉语习惯采用一些其他的方式来表达被动态,如可用:"由""给""受""加以""把""使"等。

> The outflow valve is controlled by CPC.
> 排气活门由座舱压力控制器控制。

(2) 着重描述事物过程性质和状态的英语被动句,实际上与系表结构很相近,往往可译成汉语的判断句,即将谓语动词放在"是……的"之间的结构。

> In certain special areas (landing gear bays), the LP pipes are made of stainless steel or titanium alloy.
> 在特定区域(如起落架轮舱),低压管路是用不锈钢或钛合金制造的。

(3) 英语的被动语态有时可改译成汉语的主动语态。当原句中主语为无生命的名词,而又无由介词 by 引导的行为主体时,可将原句的主语仍译为主语,按汉语习惯表达成主动句。因为汉语中的很多情况下,表达被动的含义通常不需要加"被"字,而采用主动语态的形式,如果不顾汉语习惯,强行加上"被"字表被动,有时则会使译文看上去不像汉语。

> Escape slide inflation is automatic if the door is opened with the emergency control handle in the ARMED mode.
> 当应急控制手柄在 ARMED 模式时,一旦舱门打开,撤离滑梯会自动充气。

> To prolong the duration of the oxygen supply, the oxygen is automatically diluted in the regulator with suitable amount of atmospheric air.
> 为了延长供氧时间,氧气在调节器内自动与适量的外界空气混合进行稀释。

对于有些被动句翻译时,可将原主语译成宾语,而把语言行为主体或相当于行为主体的介词宾语译成主语。

> The aircraft and the cabin pressure condition are sensed by a double pressure transducer, fixed to front face by four screws and equipped with two rubber tubes ducting the pressure from adapters (380) and (390).
> 用来感受飞机外部和客舱内部压力情况的一个双路压力传感器,由 4 个螺钉固定在前面板,并通过两根橡胶管从转接头 380 和 390 处引来压力。

有些原句中没有行为主体,翻译时可添加适当的主语,使译文通顺流畅,如人们、有人、大家、我们等。

> A few decades ago, it was thought unbelievable that the computer could have so high speed as well as small volume.
> 几十年前人们还认为,计算机能具有如此高的运行速度和如此小的体积是一件难以置信的事情。

(4) 当不需要或无法讲出动作的发出者时,如表明观点、要求、态度的被动句或描述某地发生、存在、消失了某事物的被动句,可将原句译为汉语的无主句,把原句的主语译作宾语。

> What kind of method is needed to decrease the maintenance error of AMT?
> 需要采用什么方法才能减少维修人员的差错?

UNIT 3

AIRCRAFT DESIGN BASIS

[1] Large transports are currently manufactured by three fiercely competitive companies: Boeing in the US, Airbus in Europe, and Commercial Aircraft Corporation of China Limited (COMAC), established on May 11, 2008 in Shanghai, China. COMAC is the main institution responsible for the implementation of China's major national large-scale aircraft programs, which includes medium and large passenger aircraft, such as C909, C919, and C929. Any new airplane designs must offer an advantage over the products currently produced by these manufacturers (known as "airframers").

[2] In picking the basis for a new aircraft design, the manufacturer defines the airplane in terms of range, payload, cruise speed, takeoff and landing distance. These are selected based on marketing studies and in consultation with potential customers. Two examples of decisions that need to be made are aircraft size and speed.

[3] The starting point for any vehicle system design work is to have information about current systems. Here we will use twelve recent transports as examples of current designs. We have divided them into three categories: narrow body transports which have a single aisle; wide-body aircraft which have two aisles, and regional jets which are small narrow body aircraft.

[4] Table 3-1 provides a summary of the key characteristics of these airplanes. The values shown are the design values, and the range and payload, and associated takeoff and landing distances can vary significantly. Detailed performance data can be found for these airplanes on their webs.

Table 3-1 Key current transport aircraft

Aircraft	MTOW/ lb	Empty Weight/ lb	Wingspan/ ft	No. of Pax.	Range/ nm	Cruise Mach	Takeoff distance/ ft	Landing distance/ ft
Regional Jets								
A220	138,986	N/A	115.2	135	3,350	0.78	4,000	5,000
ERJ190	114,100	63,215	94.2	98	2,000	0.82	7,562	4,258
C909	89,287	55,016	89.53	90	1,200	0.78-0.8	6,234	5,413
Narrow Body								
A320	169,800	92,000	111.8	150	3,500	0.78	5,900	4,800

续表

Aircraft	MTOW/ lb	Empty Weight/ lb	Wingspan/ ft	No. of Pax.	Range/ nm	Cruise Mach	Takeoff distance/ ft	Landing distance/ ft
B737-700	174,000	8,100	112.6	149	3,441	0.781	5,900	4,500
C919	159,700	92,700	117.5	158	3,000	0.78-0.8	7,200	5,250
Wide Body								
A330	513,670	274,650	197.8	440	6,450	0.82	8,700	5,873
A350-1000	584,400	N/A	212.4	369	8,000	0.85	N/A	N/A
A380	1,234,600	611,000	261.8	555	9,200	0.85	9,350	6,200
B747-8	973,600	420,500	224.4	364	8,000	0.75	10,170	6,560
B777-300	660,000	342,900	199.9	368	6,854	0.84	12,150	6,050
B787-8	502,200	260,000	197.2	223	7,830	0.85	7,480	4,450

[5] Aircraft flies by exploiting the laws of nature. Essentially, lift produced by the wing has to equal the weight of the airplane, and the thrust of the engines must counter the drag. The goal is to use physics principles to achieve efficient flight. A successful design requires the careful integration of a number of different disciplines. To understand the basic issues, we need to establish the terminology and fundamentals associated with the key flight disciplines. These include: aerodynamics, propulsion, control and stability, structures/materials, avionics and systems.

[6] **Aerodynamics**: The airplane must generate enough lift to support its weight, with a low drag so that the L/D ratio is "high". For a long range transport this ratio should approach 20. The lift also has to be distributed around the center of gravity so that the longitudinal (pitching) moment about the center of gravity can be set to zero through the use of controls without causing extra drag. This requirement is referred to as "trim". Extra drag arising from this requirement is "trim drag".

[7] Drag arises from two primary sources, and the viscosity of the air causes friction on the surface exposed to the airstream. This "wetted" area should be held to a minimum. To account for other drag associated with surface irregularities, the drag includes contributions from the various antennas, fairings and manufacturing gaps. Taken together, this drag is generally known as the parasite drag. The other major contribution to drag arises from the physics of the generation of lift, and is thus known as drag due to lift. When the wing generates lift, the flow field is deflected down, causing an induced angle over the wing. This induced angle leads to an induced drag. The size of the induced angle depends on the span loading of the wing, and can be reduced if the span of the wing is large. The other contribution to drag arises due to the presence of shock waves. Shock waves start to appear as the plane's speed approaches the speed of sound, and the sudden increase in drag once caused an engineer to describe this "drag rise" as a "sound barrier".

[8] **Propulsion**: Virtually all modern transport aircraft use high bypass ratio

turbofan engines. These engines are much quieter and more fuel efficient than the original turbojet engines. The turbofan engine has a core flow that passes through a compressor, then enters the combustor and drives a turbine. This is known as the hot airstream. The turbine also drives a compressor that accelerates a large mass of air that doesn't pass through the combustor, and is known as the cold flow. The ratio of the cold air to the hot air is the bypass ratio. From an airplane design standpoint, the key considerations are the engine weight per pound of thrust and the fuel consumption.

[9] Values for thrust and fuel flow of an engine are quoted for sea level static conditions. Both the maximum thrust and fuel flow vary with speed and altitude. In general, the thrust decreases with altitude, and with speed at sea level, but remains roughly constant with speed at altitude. The specific fuel consumption (sfc) increases with speed, and decreases with altitude.

[10] **Control and stability**: Safety plays a key role in defining the requirements for ensuring that the airplane is controllable in all flight conditions. Stability of the motion is obtained either through the basic airframe stability characteristics or by the use of an electronic control system providing apparent stability to the pilot or autopilot. Originally airplane controls used simple cable systems to move the surfaces. When airplanes became large and fast, the control forces using these types of controls became too large for the pilots to be able to move surfaces, and hydraulic systems were incorporated. Now some airplanes are using electric actuation. Traditionally controls were required to pitch, roll and yaw the airplane. Pitch stability is provided by the horizontal stabilizer, which has an elevator for control. Similarly, directional stability is provided by the vertical stabilizer, which incorporated a rudder for directional control. Roll control is provided by ailerons, which are located on the wings of the airplane. In some cases, one control surface may be required to perform several functions, and in some cases, multiple surfaces are used simultaneously to achieve the desired control.

[11] **Structures/materials**: Aluminum has been the primary material used in commercial transports. However, composite materials have now reached a stage of development that allows them to be widely used, providing the required strength at a much lighter weight. The structure is designed for an extremely wide range of loads, including taxiing and ground handling (bump, touchdown, etc.) and flight loads for both sustained maneuvers and gusts.

[12] Typically, transport aircraft consists of a constant cross section pressurized fuselage that is essentially round, and a wing that is essentially a cantilever beam. The constant cross section of the fuselage allows the airplane to be stretched to various sizes by adding additional frames, some in front of, and some behind the wing to allow the plane to be properly balanced. However, if the airplane becomes too long, then the tail will scrape on the ground when the airplane rotates for takeoff. The wing typically consists of spars running along the length of the wing, and ribs running between the

front and rear of the wing. The wing is designed so that fuel is carried between the front and rear spars. Fuel is also carried in the fuselage where the wing carry-through structure is located. Carrying fuel in the wing, as well as the wing support of pylon mounted engines helps reduce the structural weight required by counteracting the load due to the wing lift. Because the wing is a type of cantilever beam, the wing weight is reduced by increasing the depth of the beam, which increases the so-called thickness-to-chord ratio, t/c. This increases the aerodynamic drag. Thus, the proper choice of t/c requires a system-level tradeoff.

[13] **Avionics and systems**: Modern aircraft incorporates many sophisticated systems to allow them to operate efficiently and safely. The electronic systems are constantly changing. Advances in the various systems allowed modern transport airplanes to use two-man crews. The basic systems are:

(1) Avionics systems: communications, navigation, radar, auto-pilot, flight control system;

(2) Other Systems: air conditioning and pressurization, anti-icing, electrical power system, hydraulic system, fuel system, APU (the auxiliary power unit), landing gear.

[14] We can translate these desirable properties into specific aerodynamic characteristics. Essentially, they can be given as:

(1) Design for Performance
- Reduce minimum drag:
  — Minimize the wetted area to reduce skin friction
  — Streamline to reduce flow separation (pressure drag)
  — Distribute area smoothly, especially for supersonic a/c
  — Consider laminar flow
  — Emphasize clean design/manufacture with few protuberances, steps or gaps
- Reduce drag due to lift:
  — Maximize span (must be traded against wing weight)
  — Tailor span load to get good span, e (twist)
  — Distribute lifting load longitudinally to reduce wave drag due to lift
  — Camber as well as twist to integrate airfoil, maintain good 2D characteristics
- Key constraints:
  — At cruise: buffet and overspeed constraints on the wing
  — Adequate high lift for field performance (simpler is cheaper)
  — Alpha tail scrape. C_L goes down with sweep

(2) Design for Handling Qualities
- Adequate control power is essential:
  — Nose up pitching moment for stable vehicles
  — Nose down pitching moment for unstable vehicles
  — Yawing moment, especially for flying wings and fighters at Hi-α

— Consider the full range of CG
- Implies: must balance the configuration around the CG properly

[15] **Current typical design process**: The airplane design process is fairly well established. It starts with a conceptual design stage, where a few engineers use the sizing approaches slightly more elaborate to investigate new concepts. Engine manufacturers also provide information on new engine possibilities or respond to requests from the airframer. If the design looks promising it progresses to the next stage: preliminary design. At this point the characteristics of the airplane are defined and offered to customers. Since the manufacturer cannot afford to build the airplane without a customer, various performance guarantees are made, even though the airplane has not been built yet. This is risky. If the guarantees are too conservative you lose the sale to the competition. If the guarantees are too optimistic, a heavy penalty will be incurred.

[16] If the airplane is actually going to be built, it progresses to detail design. The following describes the progression:

- Conceptual Design (1% of the people):

  — competing concepts evaluated　　　What drives the design?
  — performance goals established　　　Will it work/meet requirement?
  — preferred concept selected　　　　 What does it look like?

- Preliminary Design (9% of the people):

  — refined sizing of preferred concept　　Start using big codes
  — design examined/establish confidence　Do some wind tunnel tests
  — some changed allowed　　　　　　　Make actual cost estimate
  　　　　　　　　　　　　　　　　　　　(you bet your company)

- Detail Design (90% of the people):

  — final detail design　　　　　　　　Certification process
  — drawings released　　　　　　　　Component/systems tests
  — detailed performance　　　　　　　Manufacturing (earlier now)
  — only "tweaking" of design allowed　Fight control system design

NEW WORDS AND EXPRESSIONS

(1) regional jet　　　　　　　　　　　　　　支线飞机
(2) manufacturer[mænjuˈfæktʃərə]　　　　　　制造商
(3) airframer[eə(r)ˈfreɪmə]　　　　　　　　　飞机制造商
(4) payload[ˈpeɪləʊd]　　　　　　　　　　　　有效载荷,商载
(5) TOGA: take off and go around　　　　　　起飞和复飞
(6) wingspan[ˈwɪŋspæn]　　　　　　　　　　　翼展
(7) aisle[aɪl]　　　　　　　　　　　　　　　通道;过道

(8) exploiting [ɪkˈsplɔɪt] 利用
(9) discipline [ˈdɪsəplɪn] 准则
(10) terminology [ˌtɜːmɪˈnɒlədʒɪ] 术语
(11) fundamental [ˌfʌndəˈmentl] 基本原理，基本法则；基础的
(12) propulsion [prəˈpʌlʃn] 推进，推进技术
(13) longitudinal [ˌlɒŋgɪˈtjuːdɪnl] 纵向的
(14) trim [trɪm] 配平
(15) antenna [ænˈtenə] 天线
(16) gap [gæp] 缝隙
(17) span loading 翼展载荷
(18) shock wave 激波
(19) turbofan [ˈtɜːbəʊfæn] 涡轮风扇式发动机
(20) turbojet [ˈtɜːbəʊdʒet] 涡轮喷气式发动机
(21) fuel [ˈfjuːəl] 燃料，燃油
(22) consumption [kənˈsʌmpʃn] 消耗，花费
(23) sea level 海平面
(24) pilot [ˈpaɪlət] 驾驶员，飞行员
(25) vertical stabilizer 垂直安定面
(26) commercial [kəˈmɜːʃl] 商用的
(27) composite material 复合材料
(28) taxi [ˈtæksɪ] 滑行
(29) handling [ˈhændlɪŋ] 操纵
(30) maneuver [məˈnuːvə] 机动
(31) frame [freɪm] 框，隔框
(32) scrape [skreɪp] 刮痕，擦
(33) spar [spɑː(r)] 翼梁
(34) rib [rɪb] 肋
(35) counteract [ˌkaʊntərˈækt] 抵抗
(36) laminar flow 层流
(37) protuberance [prəˈtjuːbərəns] 突起，结节
(38) configuration [kənˌfɪgəˈreɪʃn] 布局，构造
(39) guarantee [ˌgærənˈtiː] 保证，担保
(40) incur [ɪnˈkɜː(r)] 招致，遭受
(41) wind tunnel 风洞
(42) certification [ˌsɜːtɪfɪˈkeɪʃn] 合格审定
(43) tweak [twiːk] 扭，拧

EXERCISE

1. Translate the followings into Chinese.

(1) **Manufacturing**. The process in which raw materials are changed into finished and usable products. An aircraft is manufactured when the raw aluminum, iron, copper, and organic chemicals are changed into the finished product.

(2) **Aerodynamics**. The branch of science that deals with the forces produced by air flowing over specially shaped surfaces called airfoils. Wings and helicopter rotors produce a vertical aerodynamic force, and propellers produce a horizontal force. Aerodynamic forces inside a turbojet engine produce pressure and velocity changes in the air as it passes through the compressor and turbine.

(3) **Shock wave**. A pressure wave formed in the air as an object, such as a flight vehicle, passes through the air at a speed greater than the speed at which sound can travel. As the vehicle moves through the air, it creates disturbances, and sound waves spread out in all directions from the disturbance. Since the vehicle is flying faster than the sound waves are moving, they build up and form a shock wave at the front and the rear of the vehicle. As air passes through a shock wave, it slows down and its static pressure increases. The energy in the air is decreased. It can be divided into oblique shock wave and normal shock wave.

(4) **Sea level**. The average level of the surface of the sea. Sea level is the mid position between the level of the water at low tide and the level at high tide.

(5) **Airworthiness certificate**. A certificate issued by the civil aviation agency to the aircraft which meet the minimum standards for airworthiness as specified in the appropriate part of the civil aviation regulations.

(6) **Wind tunnel testing**. A method of determining the aerodynamic characteristics of an aircraft design. An accurate scale model of the aircraft is made and mounted in the wind tunnel on a series of balances, or scales. Wind of an accurately controlled speed is blown across the model, and the balances measure the amount of lift and drag produced by the model at different wind speeds and angles of attack.

(7) **Airfoil**. Any surface designed to obtain a useful reaction, or lift, from air passing over it. Airplane wings, propeller blades, and helicopter rotors are examples of airfoils.

(8) **Payload**. The amount of load carried by a vehicle over and above the load necessary for the operation of the vehicle. Payload is actually the amount of revenue-producing load an aircraft can carry.

(9) **Wing span**. The length of an aircraft wing, or the distance from one wing tip to the other.

(10) **Cantilever beam**. A beam with all of its support inside the beam itself. It uses no external struts, braces, or wires. A diving board is an example of a cantilever beam.

2. Answer the following questions in English.
(1) What phases are included during the aircraft typical design process?
(2) How to reduce the minimum drag during aircraft performance design?
(3) Drag arises from two primary sources, and what are they?
(4) How to reduce the drag due to lift?
(5) List the major airframers in the world currently.

SUPPLEMENTARY READING

[1] With the increase of computer power, new methods for carrying out the design of an aircraft have been developed. In particular, the interest is in using high fidelity computational simulations of the various disciplines at the very stages of the design process. The desire is to use the high fidelity analysis with numerical optimization tools to produce better designs. Here the high quality analysis and optimization can have an important effect on the airplane design early in the design cycle. Currently, high fidelity analyses are used only after the configuration shape has been "frozen". At that point it is extremely difficult to make significant changes. If the best tools can be used early, the risk and the design time will be reduced. Recent efforts have also focused on means of using large scale parallel processing to reduce the design cycle time. These various elements, taken all together, are generally known as multidisciplinary design optimization (MDO). One collection of papers has been published as a book on the subject, and there is a major conference on MDO every other year sponsored by the American Institute of Aeronautics and Astronautics (AIAA), International Society for Structural and Multidisciplinary Optimization (ISSMO) and other societies. Perhaps the best survey of our view of MDO for aircraft design is summarized in the paper by Giunta. We will outline the MDO process and issues based on these and other recent publications.

[2] Our current view of MDO is that high-fidelity codes can't be directly coupled into one major program. There are several reasons. Even with advanced computing the computer resources required are too large to perform an optimization with a large number of design variables. For 30 or so design variables, with perhaps one hundred constrains, hundreds of thousands of analyses of the high fidelity codes are required. In addition, the results of the analyses are invariably noisy, so that gradient-based optimizers have difficulty in producing meaningful result. In addition to the artificial

noise causing trouble, the design space is non-convex, and many local optima exist. Finally, the software integration issues are complex, and it is unlikely that major computational aerodynamic and structures codes can be combined. Thus, innovative methods are required to incorporate MDO into the early stages of airplane design.

[3] Instead of a brute-force approach, MDO should be performed using surrogates for the high-fidelity analyses. This means that for each design space should be constructed which uses a parametric model of the airplane in terms of design variable such as wing span and chords. The ranges of values of these design variables are defined, and a data base of analyses for combinations of the design variables should be constructed. Because the number of combinations will quickly become extremely large, design of experiments theory will need to be used to reduce the number of cases that need to be computed. Because these cases can be evaluated independently of each other, this process can exploit coarse grain parallel computing to speed the process. Once the database is constructed, it must be interpolated. In statistical jargon this means constructing a response surface approximation. Typically, second order polynomials are used. This process automatically filters out the noise from the analyses of the different designs. These polynomials are then used in the optimization process in place of the actual high-fidelity codes. This allows for repeated investigations of the design space with an affordable computational cost. A more thorough explanation of how to use advanced aerodynamics methods in MDO, including examples of trades between aerodynamics and structures, has been presented by Mason.

[4] Current issues of interest in MDO also include the consideration of the effects of uncertainty of computed results and efficient geometric representation of aircraft. MDO is an active research area, and will be a key to improving future aircraft design.

TOPIC 3：航空维修英语特点

航空维修英语词汇特点

1. 航空维修专业词汇

各行各业均存在本行业专业词汇，航空维修专业词汇大约有 1000 个，本书涵盖了大部分此类的词汇。

比如：slat 特指飞机前缘的缝翼，用于在低速时增加升力；fuselage 特指飞机的机身；等等。

值得注意的是，部分常用词，在特定环境下有特定意义。

如：check valve，两个单词都很常见，但组合起来却是"单向活门"，不能望文生义地译成"检查活门"。

此类词汇除平时多注意积累外，遇到不能理解或可能有歧义的词汇必须查阅专业词典或询问有经验人员，否则很容易闹出笑话。

2. 一词多义现象

在航空维修专业里，一个单词在不同的系统甚至不同的附件中，都可能有各自的意义。

例如：

clear

做动词：

（1）清除

> Clear the dust from the surface.
> 清除表面的灰尘。

（2）允许

> Clear to taxi.
> 允许滑行。

做形容词：

（1）无障碍的

> Make sure that the travel ranges of the flight control surfaces are clear before you pressurize/depressurize a hydraulic system.
> 在对液压系统增压/释压前，确保飞行操纵面行程范围内无障碍。

（2）没有，通常以 be clear of 的形式出现

> Make sure that the work area is clean and clear of tool(s) and other items.
> 确保工作区域干净且没有工具和其他物品。

（3）透明的

> Remove the clear ice.
> 清除透明冰。

其他如：

cable 基本含义为绳索。

在机械专业中：control cable　　　操纵钢索

在电气专业中：electrical cable　　电缆

cylinder 基本含义为圆柱体。

在有气体的系统中为"瓶体"，如：oxgyen cylinder　　氧气瓶

在液压系统中为"缸、作动筒"，如：hydraulic cylinder　　液压作动筒、液压缸

bus 基本含义为公共汽车，引申为运载（电流，数据）的装置。

在强电系统：AC bus　　交流汇流条

在弱电系统：data bus　　数据总线

switch 做名词时意思为电门，做动词时意思为转换。

open，close：打开，关闭。

在电路中，open 为断开，close 为闭合、接通。

在油路中，open 为接通、打开，close 为关闭、关断。

这一组词在手册中很常见，却是很容易搞错的词汇，而万一用错，就可能造成严重后果，甚至导致人员伤亡，一定要特别加以注意。

对于这种情况，必须认真分析并结合上下文，谨慎判断该词在本句中所要表达的具体含义。

3. 多词近义

由于英语和汉语词汇的内涵和外延不能完全重合，经常会出现多词近义现象，因此需要认真区分其细微差异。

例如：

indicator，annunciator，legend，display 都有显示、指示的意思，其区别如下：

indicator（indication）：指示器（指示），显示数值，数字式或传统表盘式。

annunciator（annunciation）：通告灯（通告牌），显示一串彩色字符，提示某种状态或故障。

legend：字符灯，显示 ON、OFF、FAULT 等彩色字符，简单提示某种状态或故障。

display：显示，显示器，通常指一块显示器或其上显示的内容。

其他如：

oil：主要用于发动机及其附件的冷却和润滑，为液态，通译为滑油。

lubricant：用于各种机械和附件的润滑和保养，通常为液态，通译为润滑剂。

grease：用于各种机械和附件的润滑和保养，通常为固态，通译为润滑脂。

实际工作中遇到类似问题，应根据上下文做出判断，如必要应咨询有经验人员，或上飞机实地考察。

UNIT 4

TYPES AND APPLICATIONS OF AEROSPACE MATERIALS

[1] **Main types of aerospace materials:** Aerospace materials are defined here as structural materials that carry the loads exerted on the safety-critical airframe during flight operations (including taxiing, take-off, cruising and landing). The materials used in the main components of engines, such as the turbine blades, are important to the safety and performance of aircraft and therefore are considered as structural materials. They have an impact through the entire life cycle of aircraft, from the initial design phase through to manufacture and certification of the aircraft, to flight operations and maintenance and, finally, to disposal at the end-of-life.

[2] An extraordinarily large number and wide variety of materials are available to aerospace engineers to construct aircraft. It is estimated that there are more than 120,000 materials from which an aerospace engineer can choose the materials for the airframe and engine. This includes many types of metals (over 65,000), plastics (over 15,000), ceramics (over 10,000), composites, and natural substances such as wood. The number is growing at a fast pace as new materials are developed with unique or improved properties.

[3] The great majority of materials, however, lack one or more of the essential properties required for aerospace structural or engine applications. Most materials are too expensive, heavy or soft or they lack sufficient corrosion resistance, fracture toughness or some other important property. Materials used in aerospace structures and engines must have a combination of essential properties that few materials possess. The demand on aerospace materials to be lightweight, structurally efficient, damage tolerant, and durable while being cost-effective and easy to manufacture rules out the great majority for aerospace applications. Other demands on aerospace materials are emerging as important future issues. These demands include the use of renewable materials produced with environmentally friendly processes and materials that can be fully recycled at the end of the aircraft life. Sustainable materials that have little or no impact on the environment when produced, and also reduce the environmental impact of the aircraft by lowering fuel burn (usually through reduced weight), will become more important in the future.

[4] It is estimated that less than about one hundred types of metal alloys, composites, polymers and ceramics have the combination of essential properties needed for aerospace

applications. The main groups of materials used in aerospace structures are aluminum alloys, titanium alloys, steels, ceramics and composites. In addition to these materials, superalloys are important structural materials for engines.

[5] **Aluminum alloy**: Aluminum alloy is used extensively for several reasons, including their moderately low cost; ease of fabrication which allows it to be shaped and machined into structural components with complex shapes; light weight; and good stiffness, strength and fracture toughness. Similarly to any other aerospace material, there are several problems with using aluminum alloys, and these include susceptibility to damage by corrosion and fatigue.

[6] There are many types of aluminum used in aircraft whose properties are controlled by their alloy composition and heat treatment. The properties of aluminum are tailored for specific structural applications; for example, high-strength aluminum alloys are used in the upper wing skins to support high bending loads during flight whereas other types of aluminum alloys are used on the lower wing skins to provide high fatigue resistance.

[7] High-strength aluminum alloy is the most used material for the fuselage, wing and supporting structures of many commercial airliners and military aircraft, particularly those built before the year 2000. In recent years the use of aluminum has fallen owing to the growing use of fiber-polymer composite materials. The competition between the use of aluminum and composite is intense, although aluminum will remain an important aerospace structural material.

[8] **Titanium alloy**: Titanium was first used in military and commercial aircraft during the 1950s. The original need for titanium arose from the development of supersonic aircraft capable of speeds in excess of Mach 2. The skins of these aircraft require heat-resistant materials that do not soften owing to frictional heating effects at supersonic speeds. Conventional aluminum alloys soften when the aircraft speed exceeds about Mach 1.5 whereas titanium remains unaffected until Mach 4-5.

[9] The use of titanium in commercial aircraft has increased over recent decades, albeit slowly owing to the high cost of titanium metal and the high costs of manufacturing and machining titanium components. Although the mechanical properties of titanium are better than those of aluminum, the material and manufacturing costs are much higher and it is uneconomical to use in structural components unless they need to be designed for high loads. For this reason, the structural weight of titanium in airliners is typically under 10%, although higher amounts are used in new aircraft types such as the A350 and B787.

[10] Titanium alloys are used in both airframe structures and jet engine components because of their moderate weight, high structural properties (e.g. stiffness, strength, toughness, fatigue), excellent corrosion resistance, and the ability to retain their mechanical properties at high temperature. Various types of titanium alloys with different compositions are used, although the most common is Ti-6Al-4V which is used in both aircraft structures and engines. Engine components made of titanium include fan blades, low-pressure compressor parts, and plug and nozzle assemblies in the exhaust section.

[11] **Steel**: Steel is the most commonly used metal in structural engineering, however its use

as a structural material in aircraft is small (under 5%-10% by weight). The steels used in aircraft are alloyed and heat-treated for very high strength, and are about three times stronger than aluminum and twice as strong as titanium. Steels also have high elastic modulus (three times stiffer than aluminum) together with good fatigue resistance and fracture toughness. However, steel is not used in large quantities for several reasons, with the most important being its high density, nearly three times as dense as aluminum and over 50% denser than titanium. Other problems include the susceptibility of some grades of high strength steel to corrosion and embrittlement which can cause cracking.

[12] This combination of properties makes steel a material of choice for safety-critical structural components that require very high strength and where space is limited, such as the landing gear and wing box components.

[13] **Superalloys:** Superalloys are a group of nickel, iron-nickel and cobalt alloys used in jet engines. These metals have excellent heat resistant properties and retain their stiffness, strength, toughness and dimensional stability at temperatures much higher than the other aerospace structural materials. Superalloys also have good resistance against corrosion and oxidation when used at high temperatures in jet engines. The most important type of superalloy is the nickel-based material that contains a high concentration of chromium, iron, titanium, cobalt and other alloying elements. Superalloys are used in engine components such as the high-pressure turbine blades, discs, combustion chamber, afterburners and thrust reversers.

[14] **Ceramics:** Ceramics often consist of (metal) oxides and metals, in which ionic bonds between the different atoms provide the material structure. Ceramics are not suitable for structures. They are too brittle and have poor processing features.

[15] However, they are applied in some space applications, for instance for thermal protection of the metallic or composite structure. Turbine blades contain rows of hollow aerofoils for cooling to increase the engine operating temperature. Cool air flows through the holes, which are located just below the surface, to remove heat from the superalloy. The aerofoils are remarkably effective at cooling, which allows increased operating temperature and associated improvements in engine efficiency. To increase the operating temperature even further, the hottest engine parts are coated with a thin ceramic film to reduce heat flow into the superalloy. The film is called a thermal barrier coating, which has higher thermal stability and lower thermal conductivity than nickel superalloy. The use of the coating allows higher operating temperatures in the turbine section. The coating provides heat insulation and this lowers the temperature of the superalloy engine component. Thermal barrier coatings can survive temperatures well in excess of the melting temperature of the superalloy itself, and also provide protection from the effects of thermal fatigue and creep and the oxidizing effect of sulfates and other oxygen-containing compounds in the combustion gases. Thermal barrier coatings are used on engine components in the combustion chamber and turbine sections, including high-pressure blades and nozzle guide vanes.

[16] **Composites:** Composites are an important group of materials made from a mixture of

metals, ceramics and/or polymers to give a combination of properties that cannot otherwise be achieved by each of their own. Composites are usually made by combining a stiff, strong but brittle material with a ductile material to create a two-phase material characterized by high stiffness, strength and ductility. Composites consist of a reinforcement phase and a matrix phase and, usually, the reinforcement is the stiffer, stronger material which is embedded in the more ductile matrix material. For example, carbon-epoxy composite used in aircraft structures is a mixture of carbon fibers (the reinforcement phase) embedded in epoxy resin (the matrix phase).

[17] Composites are usually classified according to the material used for the matrix: metal matrix, ceramic matrix or polymer matrix. There is also a special type of composite called fiber-metal laminate that does not really fit into any of these three classes. The composites used in aircraft are almost exclusively polymer matrix materials, with the polymer often being a thermoset resin (e.g. epoxy, bismaleimide) and occasionally a high-performance thermoplastic (e.g. PEEK).

[18] There are many advantages as well as several problems with using carbon fiber composites rather than aluminum in aircraft. The advantages include reduced weight, capability to manufacture integrated structures from fewer parts, higher structural efficiency (e.g. stiffness/weight and strength/weight), better resistance against fatigue and corrosion, radar absorption properties, good thermal insulation, and lower coefficient of thermal expansion. The disadvantages of composites include higher cost, slower manufacturing processes, anisotropic properties making design more difficult, low through thickness mechanical properties and impact damage resistance, higher sensitivity to geometric stress raisers such as notches, lower temperature operating limit, and lower electrical conductivity.

[19] Carbon fiber composites are used in the major structures of aircraft, including the wings, fuselage, empennage and control surfaces (e.g. rudder, elevators, ailerons). Composites are also used in the cooler sections of jet engines, such as the inlet fan blades, to reduce weight. In addition to carbon fiber composites, composites containing glass fibers are used in radomes and semi structural components such as fairings and composites containing aramid fibers are used in components requiring high impact resistance.

NEW WORDS AND EXPRESSIONS

(1) extraordinarily[ɪkˈstrɔːdnrəlɪ]　　　极其,极端地;奇怪地
(2) ceramic[səˈræmɪk]　　　陶瓷;陶瓷制品
(3) composite[kəmˈpəʊzɪt]　　　复合材料
(4) fracture[ˈfræktʃə(r)]　　　破裂,使破裂
(5) toughness[ˈtʌfnəs]　　　韧性

(6) polymer ['pɔlɪmə] 聚合物
(7) stiffness ['stɪfnəs] 刚度
(8) susceptibility [sə,septə'bɪlətɪ] 敏感性；灵敏度；磁化系数
(9) albeit [,ɔ:l'bi:ɪt] 虽然，尽管
(10) elastic modulus 弹性模量
(11) embrittlement [ɪm'brɪtəlmənt] 脆化，脆裂
(12) cobalt ['kəubɔ:lt] 钴
(13) oxidation [,ɔksɪ'deɪʃn] 氧化
(14) concentration [,kɔnsn'treɪʃn] 浓度；集中
(15) atom ['ætəm] 原子
(16) brittle ['brɪtl] 易碎的，脆弱的
(17) conductivity [,kɔndʌk'tɪvətɪ] 导电性
(18) sulfate ['sʌlfeɪt] 硫酸盐
(19) matrix ['meɪtrɪks] 基体；矩阵
(20) epoxy resin 环氧树脂
(21) thermoset ['θə:məuset] 热固性
(22) bismaleimide ['bɪsmæli:'maɪd] 双马来酰亚胺
(23) thermoplastic [,θə:məu'plæstɪk] 热塑性塑料
(24) coefficient [,kəuɪ'fɪʃnt] 系数
(25) anisotropic [æn,aɪsəu'trɔpɪk] 各向异性的
(26) notch ['nɔtʃ] 切口，缺口，凹口
(27) semi-structural 半结构的
(28) aramid fibre 芳纶纤维

EXERCISE

1. Translate the followings into Chinese.

(1) **Airframe**. The fuselage, booms, nacelles, cowlings, fairings, airfoil surfaces (including rotors but excluding propellers and rotating airfoils of engines), and landing gear of an aircraft and their accessories and controls.

(2) **Ceramic**. Any of several hard, brittle, heat-resistant, noncorrosive materials made by shaping and then firing a mineral, such as clay, at a high temperature. Ceramic electrical insulators are stronger than glass, and they withstand high temperatures better than glass.

(3) **Composite**. Something made up of different materials combined in such a way that the characteristics of the resulting material are different from those of any of the components.

(4) **Corrosion**. An electrolytic action which takes place inside a metal or on its

surface. The metal reacts with an electrolyte, and part of the metal is changed into a salt, which is the corrosion. Corrosion is dry and powdery and has no physical strength.

(5) **Alloy**. A physical mixture of chemical elements combined with a metal to change its characteristics.

(6) **Fatigue**. A condition existing in a metal which causes it to lose some of its strength. Fatigue occurs when the metal is subjected to a series of stress reversals, such as happens when the metal is repeatedly bent back and forth.

(7) **Atom**. The smallest particle of a chemical element that can exist, either alone or in combination with other atoms. An atom is made up of a nucleus, which contains protons and neutrons and electrons, which spin around the nucleus. In a balanced atom, there are as many electrons spinning around the nucleus as there are protons in the nucleus.

(8) **Thermoplastic material**. A type of plastic material that becomes soft when it is heated, and hardens when it is cooled. Thermoplastic materials can be softened and hardened many times without the strength or quality of the material being affected. Transparent acrylic plastic resin is an example of a thermoplastic material.

(9) **Stress**. A force within an object that tries to prevent any outside force changing its shape.

(10) **Stiffener**. A piece of formed sheet metal or extrusion riveted to a larger piece of thin sheet metal to give it rigidity and stiffness. Some types of aircraft sheet metal structure have large areas of thin sheet metal that vibrate and cause noise and weaken the metal. To prevent this metal from vibrating, a stiffener in the shape of a hat section, a channel, or an angle is riveted across the sheet. This stiffens it enough to stop the vibration.

2. Answer the following questions in English.

(1) What are the aerospace materials used on the current aircraft?

(2) What are the characteristics of the Aluminum alloys material?

(3) What are the characteristics of the Titanium alloys?

(4) There are many advantages as well as several problems with using carbon fiber composites rather than aluminum in aircraft, what are they?

(5) Which part(s) of aircraft are made of composite materials?

SUPPLEMENTARY READING

[1] The future success of the aerospace industry both in terms of the cost-effective manufacture of new aircraft and the cost-effective extension of the operating life of

existing aircraft is reliant on on-going improvements to existing materials and the development of new materials. Advances in materials technology is classified as evolutionary or revolutionary. Evolutionary advances mean that small, incremental improvements are made to existing materials, such as a new alloy composition, processing method or heat treatment. Examples include the addition of new alloying elements to nickel superalloys to increase the creep resistance and maximum operating temperature or the development of a new thermal ageing treatment for aluminum alloys to increase their resistance to stress corrosion. The evolutionary approach is often preferred by the aerospace industry since past experience has shown that almost every new material has some initial problems. The aerospace industry is more comfortable with incremental improvements on conventional materials, for which they have good knowledge of the design, manufacture, maintenance and repair issues.

[2] Revolutionary advances are the application of new materials to structures or engines that are different to previously used materials. An example was the first-time that carbon-fiber composite was used to fabricate a primary load-bearing structure (the tail section) on a commercial airliner (B777) in the mid-1990s. Another example was the use of glass reinforced aluminum laminates (GLARE) in the A380 in 2005, which was the first application of a fiber-metal laminate in an aircraft fuselage. Revolutionary materials usually have had limited success in being directly incorporated into aircraft owing to the high costs of manufacturing, qualification and certification. The cost and time associated with developing a new material and then testing and certifying its use in safety-critical components can cost an aerospace company hundreds of millions of dollars and take 5-10 years or longer. The introduction of new material can require major changes to the production infrastructure of aircraft manufacturing plants as well as to the in-service maintenance and repair facilities. For example, some suppliers of structural components to the Boeing 787 had to make major changes to their production facilities from metal to composite manufacturing, which required new design methods, manufacturing processes and quality control procedures as well as reskilling and retraining of production staff. Despite the challenges, revolutionary materials are introduced into new aircraft when the benefits outweigh the potential problems and risks.

[3] Evolutionary and revolutionary advances across a broad range of materials technologies are on-going for next-generation aircraft. It is virtually impossible to give a complete account of these advances because they are too numerous. Some important examples are: development of high-temperature polymers for composites capable of operating at 400°C or higher; new polymer composites that are strengthened and toughened by the addition of nanosized clay particles or carbon nanotubes; new damage tolerant composites that are reinforced in the through-thickness direction by techniques such as stitching, orthogonal weaving or z-pinning; multifunctional materials that serve

several purposes such as thermal management, load-bearing strength, self-assessment and health monitoring, and self-actuation functions; bioinspired polymer materials with self-healing capabilities; sandwich materials that contain high-performance metal-foam cores, truss or periodic open-cell cores; new tough ceramic materials with improved structural capabilities; rapidly solidified amorphous metals with improved mechanical properties and corrosion resistance; and new welding and joining processes for dissimilar materials. The aerospace materials for this century are sure to be just as ground-breaking and innovative as the materials used in the past century. On-going improvement in structural and engine materials is essential for the advancement of aerospace engineering. After many years of commercial service it might be expected that structural and engine material technology would be approaching a plateau, and the pace of innovation would decline. However, customer demands for higher performance, lower operating costs and more "environmentally friendly" propulsion systems continue to drive materials research to ever more challenging goals.

[4] The general approach to implementing new materials into aircraft is through evolutionary advances, which simply means incremental improvements to the properties of the materials already in use. Revolutionary advances which involve the application of new material are less common, although they do occur when the benefits outweigh the cost and risk.

TOPIC 4：航空维修英语特点（续）

航空维修英语文句特点

1. 结构简单，很少有时态、语态变化

> Put a warning notice in the following position to tell persons not to operate these pushbutton switches.
> 在以下位置放置警告标牌以告知他人不要操作这些按钮电门。

2. 祈使句居多

> Cut a piece of the tape of 25 mm (0.9842 in)×40 mm (1.5748 in) and put it in position in the area of the panel fastener.
> 剪下一片25毫米(0.9842英寸)×40毫米(1.5748英寸)的胶带，并把它粘在面板紧固件区域。

3. 格式标准化

> Make sure that the work area is clean and clear of tool(s) and other items.
> 确保工作区域干净且没有工具和其他物品。

4. 空客手册的部分编写者母语非英语，手册中时有语言不规范之处

> Open, safety and tag the following circuit breaker(s).
> safety 本为名词，但在本句中做动词使用，意为打保险。
> 三个动词共用一个宾语，此句直译为：
> 将下列电路跳开关断开，打保险并挂标牌。
> 在实际维修工作中，为方便理解，本句可意译为：
> 拔出下列电路跳开关，打保险并挂标牌。

5. 原文手册本身也可能有错误的地方，容易造成误解

例如：

> Remove the blanking plugs from the cabin recirculation fan (15), the sleeve (1) and from the check valve (7).
> 本句出自维护手册，根据句子结构可译为"从客舱再循环风扇(15)上拆下堵盖，并从单向活门(7)上拆下衬套(1)"。
> 本句中 and 和 from 两个单词并列让人无法理解。维修人员也认为可能是 and 的位置应放在 the sleeve 前，后来重新查看了维修手册，找到了该部件的安装图，经反复推敲，并向一线相关人员求证，最终判定，本句中的 from 为冗字，是手册编写过程中出现的错误，应去掉。正确的翻译应为：
> 从客舱再循环风扇(15)、衬套(1)和单向活门(7)上拆下堵盖。

6. 原理描述部分有复杂的句子

维修人员使用的原文手册中操作原理部分(如维护手册的系统描述(system description section,SDS)部分)和某些法规性文件存在一些长达数十字甚至上百字的复杂长句。这些句子往往比较重要,如产生误解会有一定不良后果,需要认真分析。

对复杂长句的理解不能从查生词开始,因为一些词汇含义很多,而通用字典甚至某些专业字典不能完全覆盖所需要的专业释义。

这种情况下,首先应该确定句子的主要成分(主、谓、宾语),然后判断各从句和修饰成分与主要成分的逻辑关系,再根据语言环境判断关键词的释义。经过这三个步骤,就可以基本理解句子的含义了。接下来需要将之整理成通俗易懂的汉语,才算完成对一个复杂长句的正确翻译。

例句1:

> The Maint. & Overhaul Base will not maintain or alter any article, which it is not rated, nor will it maintain or alter any article which it is rated if it requires special technical data, equipment, or facilities that are not available.
>
> 首先可以看出这是两个并列分句,以连词 nor 为分界。主语是 The Maint. & Overhaul Base,两个分句的谓语都是 will not maintain or alter,宾语都是 any article。对第一个分句,定语从句 which it is not rated,是对宾语 article 的说明;对第二个分句,除了同样有一个定语从句 which it is rated 修饰宾语外,还有一个 if 开头的条件状语从句用来限制谓语的范围。在这个条件状语从句中,又包含一个定语从句 that is not available 修饰宾语 special technical data,equipment,or facilities。
>
> 接下来再看词汇,很明显 maintain 和 alter 的意思为维修和改装,那么,作为其宾语的 article 就不可能表示"文章",而是应该指"项目"。rate,原义为划分等级,在这里是指维修站被批准的维修范围。
>
> The Maint. & Overhaul Base will not maintain or alter any article, for which it is not rated,
> 维修基地不维修或改装任何项目, 不是被评定范围内的
> nor will it maintain or alter any article for which it is rated if it requires special technical data,
> 也不 维修或改装 任何项目 是被评定范围内的 如果它需要专用资料
> equipment, or facilities that are not available.
> 设备 或 设施 没有
>
> 现在整个句子基本上可以翻译为:
> 维修基地不会维修或改装不在其许可维修范围内的任何项目,也不会在如果需要现在没有的专用资料、设备或设施的情况下,维修或改装在其许可维修范围内任何项目。
>
> 显然这样的直译不符合汉语习惯,可以将它再整理一下,使之通顺易懂:
> 维修基地不会维修或改装不在其许可维修范围内的任何项目,也不会维修或改装在其许可维修范围内,但所需的专用资料、设备或设施不具备的任何项目。

例句2：

ALL POINTS, WHEN NOT IN USE, HAVE HOIST POINT PLUGS INSTALLED WHICH FIT FLUSH WITH THE NACELLE SURFACE AND IN ADDITION, PREVENT FOREIGN OBJECTS FROM ENTERING AND LODGING IN THE HOIST POINT OPENINGS.

在维护手册中，所有警告和告诫信息均为大写，用来强调这些信息的重要性。如果不遵守这些信息，会导致人员受伤或设备损坏。因此，能够迅速准确阅读全文大写的句子也是机务维修人员的基本功之一。

本句中，主语为 ALL POINTS，谓语 HAVE，宾语 HOIST POINT PLUGS，WHEN NOT IN USE 是时间状语从句用于限定谓语，定语 INSTALLED 和定语从句 WHICH FIT FLUSH WITH THE NACELLE SURFACE IN ADDITION, PREVENT FOREIGN OBJECTS FROM ENTERING AND LODGING IN THE HOIST POINT OPENINGS 用来修饰宾语。定语从句本身也由两个分句构成。

HOIST POINT 意为起吊点；PLUG 堵塞；FLUSH 在这里是副词，译为平齐地；NACELLE 为发动机短舱；LODGING 本意为容纳、存放，在这里根据前面的名词 FOREIGN OBJECTS（外来物），可译为残留。

ALL POINTS, WHEN NOT IN USE, HAVE HOIST POINT PLUGS INSTALLED
所有吊点，　　　在不使用的时候　　　有起吊点堵盖　　　安装的
WHICH FIT FLUSH WITH THE NACELLE SURFACE AND IN ADDITION, PREVENT
防止　　　　与短舱表面平齐　　　　且　　　　也
FOREIGN OBJECTS FROM ENTERING AND LODGING IN THE HOIST POINT
　外来物　　　　　　进入　　并　　残留　　　在起吊点
OPENINGS.
开口。

直译：

所有吊点在不使用时，都装有齐平于发动机短舱表面的同时能防止外来物进入并残留在吊点开口里的起吊器堵盖。

意译：

当不使用时，所有吊点都装有起吊点堵盖，它们与发动机短舱表面平齐，还能防止外来物进入并残留在吊点开口内。

UNIT 5

NDT AND CORROSION OF AEROSPACE MATERIALS

[1] **NONDESTRUCTIVE INSPECTION**: Most damage is impossible to detect by eyes because it is too small and buried below the component surface. The aerospace industry cannot rely solely on visual examination to determine material quality. Instead, the industry is reliant on nondestructive inspection (NDI), also called nondestructive testing (NDT) and nondestructive evaluation (NDE), to assess material integrity. As the name implies, NDI does not inflict damage on the component during the inspection process (as opposed to other methods which require destruction of the component to detect the damage).

[2] Whether for the inspection of as-manufactured or in-service components, NDT is used to determine the type, size and location of damage which is essential information to assess the flight-worthiness and residual strength of the structure. There are various NDI methods, and the types most often used by the aerospace industry are: ultrasonics, radiography, thermography, eddy current, magnetic particle, dye penetrant and acoustic emission. Quite often two or more methods must be used in combination to obtain a complete description of the type, size and location of internal damage.

[3] **Visual inspection**: Two of the simplest inspection methods are visual inspection and tap testing. Visual inspection involves the careful examination of the material surface with the eyes, often assisted with a magnifying glass. Although simple, visual inspection is the first step in any NDI method, and it can identify obvious signs of damage. However, visual inspection is only suitable for detecting surface damage such as large cracks or general or exfoliation corrosion in metals, and it is not suitable when the damage is buried below the surface of metal or composite components.

[4] **Tap testing**: Tap testing involves the repeated tapping over the component surface using a coin, soft hammer or some other light object to produce a ringing noise. Damage immediately below the surface may be detected using tap testing by a change in the pitch of the noise. For example, tapping the surface of damaged composite material containing large delamination cracks can produce a dull sound compared with the higher-

pitch ringing noise of the damage-free material. However, tap testing is not always reliable and can easily fail to detect damage. Instrumented tap testing devices which measure and analyze the noise generated by the tapping are available to eliminate the need for human hearing, which is not always sensitive to small changes in pitch. Both visual inspection and tap testing can be used for the initial inspection of aircraft components, but more sophisticated NDI methods are needed for reliable inspections.

[5] **Ultrasonics**: Ultrasonics is an NDI method used widely in the aerospace industry to inspect aircraft structures and engine components. Ultrasonics is used for the detection of both manufacturing defects and in-service damage. Although ultrasonics cannot detect every type of damage, it can determine the presence of common types of damage found in metals (e.g. voids, corrosion damage, fatigue cracks) and composites (e.g. delamination, porosity). Ultrasonics is best suited for the detection of planar damage such as cracks aligned parallel with the surface. The technique is not well suited to detecting damage aligned parallel with the propagation direction of the acoustic waves, although angled probes can be used.

[6] **Radiography**: Radiography involves the use of radiation, such as x-rays, γ-rays or high-speed neutrons, to detect damage in solids. Radiography can detect defects such as voids, intermetallic inclusions, corrosion damage, and cracks larger than 0.5-1.25mm, which is below the critical damage size in aircraft structures. However, the shape and orientation of the defect affect how easily it is detected. Long cracks aligned parallel with the direction of radiation are more easily detected than cracks running perpendicular to the incident radiation. Therefore, it is necessary to inspect a component with different radiation angles to ensure cracks at different orientations are detected.

[7] **Thermography**: Thermography is used by the aerospace industry for the rapid, wide-area inspection of components. Thermography can detect delamination cracks, porous regions and foreign objects in composites and intermetallic inclusions and large voids in metals. There are two main types of infrared thermography known as passive and active. Active thermography is the more widely applied of the two. The active method involves short duration heating (usually less than 1 s) of the component surface using flash tubes, hot-air guns or some other controllable heating device. The passive thermography method uses internally generated heat, often from damage growth, rather than externally applied heat. Heat generated by damage growth is measured as a hot spot on the component surface. This technique is not as popular as active thermography because the material must be damaged to generate the internal heat and, therefore, it is rarely used to inspect aircraft components.

[8] **Eddy current**: The eddy current method is widely used to inspect metallic aircraft components for surface cracks and corrosion damage. Eddy current equipment is lightweight and portable, thus allowing for inspection of grounded aircraft. However, access to both sides of the component is necessary for complete inspection. Eddy current

testing only works on conductive materials such as aerospace metal alloys. It cannot be used to inspect insulating materials with low electrical conductivity, such as fiberglass composites, and difficulties are experienced with ferromagnetic materials.

[9] **Magnetic particle**: Magnetic particle inspection is a simple NDI method used to detect cracks at the surface of ferromagnetic materials such as steels and nickel-based alloys. The inspection process begins with the magnetization of the component. The surface is then coated with small magnetic particles, which is usually a dry or wet suspension of iron filings. Surface cracks or corrosion pits create a flux leakage field in the magnetized component. The magnetic particles are attracted to the flux leakage and thereby cluster at the crack. This cluster of particles is easier to see than the actual crack, and this is the basis for magnetic particle inspection.

[10] **Liquid dye penetrant**: Liquid dye penetrant is used to locate surface cracks in materials, but cannot detect subsurface damage. The component surface is cleaned before a visible or fluorescent liquid dye is applied using a spray, brush or bath. The dye seeps into surface cracks by capillary action. Excess dye retained on the surface is wiped off leaving only the dye that has seeped into the cracks. Chemical developer is then applied and it reacts with the dye, drawing it from the crack on to the surface. The dye can then be observed, either because it changes the color of the developer or because it fluoresces under ultraviolet light. Liquid dye penetrant is a popular inspection method because it is simple, inexpensive and can detect cracks to a depth of about 2 mm. The main drawback is that the method can only detect surface breaking cracks. Despite this problem, it is often used to inspect aircraft components, particularly engine parts, which are susceptible to surface cracking.

[11] **Acoustic emission**: Acoustic emission involves the detection of defects using sounds generated by the defects themselves. The component is subjected to an applied elastic stress; usually just below the design load limit. When cracks and voids exist in materials, the stress levels immediately ahead of the defect are several times higher than the surrounding material. This is because cracks act as a stress raiser. Any plastic yielding and microcracking that occurs ahead of the defect owing to the stress concentration effect can generate acoustic stress waves before any significant damage growth. The waves are generated by the transient release of strain energy owing to microcracking. The waves are detected using sensitive acoustic transducers located at the surface. The transducers are passive, that is, they only "listen" for sounds and do not generate the acoustic waves that the transducers used in ultrasonics do. Several transducers are placed over the test surface to determine the damage location. Certain defects have a characteristic sound frequency value and this is used to determine the type of damage present in the material. For example, delamination in carbon-epoxy composites have a characteristic frequency of about 100 kHz whereas damaged fibers emit sound at around 400 kHz.

[12] **Corrosion of aerospace materials**: Corrosion is simply defined as the chemical attack of metals that results in deterioration and loss of material. A corrosive fluid is usually involved, with the most common being water containing reactive chemicals (such as chloride ions). Moisture condenses on metal surfaces and can seep into and drip down the inside surface of the fuselage and around the lavatories and galley, causing corrosion in often hard-to-access areas. The low humidity (under 5%-8%) inside pressurized aluminum fuselages helps minimize condensation, although the dry air affects the comfort of passengers. There is no single type of corrosion that occurs in aircraft. Instead, corrosion can take many forms, including: general (or uniform) surface corrosion, galvanic (or two-material) corrosion, pitting corrosion, crevice corrosion, intergranular corrosion, fretting corrosion, exfoliation corrosion, and stress corrosion.

[13] **General surface corrosion**: General surface corrosion (also called uniform surface corrosion) is the most common type of corrosion. It involves an electrochemical reaction that proceeds uniformly over the entire exposed surface of the metal. General corrosion is responsible for the greatest destruction of metal on a tonnage basis, although it is not usually a serious corrosion problem for aircraft owing to the surface protection measures. General corrosion of metals used in the airframe only occurs when the surface protection is damaged or incorrectly applied. When it does occur, general corrosion occurs at the surface and is detected during maintenance inspection by the presence of grey or white powdery deposits. These deposits are the residual solid by-product of the corrosion process.

[14] **Pitting corrosion**: Pitting is one of the most destructive and insidious types of corrosion. Without appropriate protection, metals such as aluminum, steel and magnesium used in aircraft are susceptible to pitting corrosion, which is a form of extremely localized attack that results in small holes. Pitting can start at precipitates at the surface of certain alloys when the particle has a different electrochemical potential from the surrounding metal matrix. Pitting can also occur in surface regions where the corrosion protective layer is absent. When surface protection is used small gaps in the layer can occur because of incorrect application or in-service damage by abrasion, erosion or some other event. The corrosion forms as a hole at the gap in the protective layer, which then develops into a wider cavity below the surface.

[15] **Crevice corrosion**: Crevice corrosion (also called concentration cell corrosion) is the most common type of corrosion damage found on many older aircraft which have not been adequately maintained. It is an aggressive form of corrosion that occurs locally inside crevices and other shielded areas of metals exposed to corrosive fluid. Crevice corrosion can quickly develop into pitting or exfoliation corrosion when left untreated, depending on the types of metal and corrosive fluid. It is for this reason that crevices within aircraft, such as fastener holes and joints, must be sealed with a durable protective coating that stops the ingress of corrosive fluid. Crevice corrosion is one of

the most common types of corrosion found in aircraft, and it usually occurs in crevices under fastener heads, under loose paint, within delaminated bonded joints, or in unsealed joints.

[16] **Intergranular corrosion**: Intergranular corrosion involves the localized attack of the grain boundaries in a metal. Corrosion damage occurs along the grain boundaries whereas the core of the grains is unaffected. Intergranular corrosion occurs because the electrochemical potential of the grain boundary is different from the grain core and, therefore, the grain boundary and grain core form, respectively, the anode and cathode of a tiny galvanic cell. The grain boundary is anodic because its chemical composition is different to the grain core. The difference can be caused by a higher concentration of impurity elements at the grain boundaries, depletion of alloying elements in the grain boundary region, or some other chemical difference between the boundary and core of a grain. The formation of other types of precipitates, either along the grain boundaries or inside the grains, can also cause a difference in the electrochemical potential between the outer and inner regions of grains. In the presence of a corrosive fluid, a small-scale galvanic cell is created between the core and boundary of a grain and this leads to intergranular corrosion. The corrosion process causes the gradual disintegration of the metal by grains breaking away from the surface after the boundaries have been corroded. Intergranular corrosion is a potential problem for many types of heat-treatable aluminum and magnesium alloys used in aircraft, although it may be avoided by the use of surface protection.

[17] **Exfoliation corrosion**: A special type of intergranular corrosion is called exfoliation corrosion, which involves the lifting of surface grains by the force of corrosion products at the grain boundaries. Exfoliation corrosion starts at the surface but it mainly involves subsurface attack that proceeds along narrow paths parallel with the surface. The attack is usually along the grain boundaries (intergranular corrosion). When the grain boundaries corrode they form corrosion products that exert pressure on the surface grains which forces them upwards. This causes the grains to peel back like the pages of a book. Exfoliation corrosion is characterized by leafing of thin layers of uncorroded metal between layers of corrosion product. The grains are removed from the surface by abrasion or other mechanical action, which allows the underlying grains to then lift up and the exfoliation process to continue. High-strength aluminum alloys are susceptible to exfoliation corrosion, and their resistance to this type of corrosion is improved by over-ageing during heat treatment.

[18] **Fretting corrosion**: Fretting corrosion, which is also called friction oxidation or wear oxidation among several other terms, involves the deterioration of contacting metals subjected to vibration or slip. The fretting (or rubbing) action results in fine particle fragments being abraded from one or both materials. These fragments oxidize into hard, abrasive particles which wear and destroy the metal surface. The process is

considered corrosive because the metal particles must oxidize, which is a form of dry corrosion. Scratching and abrasion of the metal by the hard oxide particles causes a loss in dimensional tolerance between contacting surfaces such as structural joints. In extreme cases, it can cause seizing and galling of moving parts. Fretting corrosion is not usually a common problem with aircraft, although it has contributed to several aircraft accidents.

[19] **Stress-corrosion cracking**: Although stress-corrosion cracking is not the most common form of corrosion, it does account for between 5% and 10% of all aircraft component failures. Stress-corrosion cracking, which is also called environmentally assisted stress corrosion, is caused by the combination of stress and corrosion. Stress-corrosion cracks often initiate at pits, notches or other stress raiser sites on the metal surface in the presence of a corrosive fluid. Stress-corrosion cracks often grow along the grain boundaries or (less often) through the grains via a brittle-type mode of fracture. A major problem with the stress-corrosion cracking process is that the cracks are difficult to detect by visual inspection of the metal surface. High-strength aluminum alloys are susceptible to stress-corrosion cracking when exposed to water and many other corrosive fluid. High-strength steel components are also susceptible to stress-corrosion cracking, and must be protected using a surface coating such as cadmium or chromium plating.

NEW WORDS AND EXPRESSIONS

(1) nondestructive testing (NDT)   无损检测
(2) nondestructive inspection (NDI)   无损探伤
(3) nondestructive evaluation (NDE)   无损评价
(4) inflict[ɪnˈflɪkt]   造成；使遭受
(5) worthiness[ˈwɜːðɪnəs]   值得；相当；有价值
(6) ultrasonics[ˌʌltrəˈsɒnɪks]   超声波
(7) radiography[ˌreɪdɪˈɒɡrəfɪ]   射线成像
(8) thermography[θəːˈmɒɡrəfi]   热成像
(9) eddy current   涡流
(10) penetrant[ˈpenətrənt]   渗透剂
(11) acoustic emission   声发射检测
(12) tap[tæp]   敲击
(13) magnifying[ˈmæɡnɪfaɪɪŋ]   放大，放大的
(14) hammer[ˈhæmə(r)]   锤子，锤击
(15) planar[ˈpleɪnə(r)]   平面的；二维的；平坦的
(16) probe[prəʊb]   探头
(17) neutron[ˈnjuːtrɒn]   中子

(18) void[vɔɪd] 砂眼,气泡;空隙;空的
(19) porous[ˈpɔːrəs] 多孔的,疏松的
(20) infrared[ˌɪnfrəˈred] 红外线,红外线的
(21) flux[flʌks] 流量,通量(磁、电)
(22) cluster[ˈklʌstə(r)] 群;簇
(23) fluorescent[fləˈresnt] 荧光,荧光的
(24) ultraviolet[ˌʌltrəˈvaɪələt] 紫外线
(25) transducer[trænzˈdjuːsə(r)] 传感器
(26) chloride ions 氯离子
(27) moisture[ˈmɔɪstʃə(r)] 水分;湿度
(28) galvanic[gælˈvænɪk] 电偶
(29) electrochemical[ɪˌlektrəʊˈkemɪkəl] 电化学的
(30) tonnage[ˈtʌnɪdʒ] 吨位,总重量
(31) deposit[dɪˈpɑːzɪt] 沉积物
(32) intergranular[ˌɪntəˈgrænjʊlə] 晶间

EXERCISE

1. Translate the followings into Chinese.

(1) **Nondestructive inspection**. Any type of inspection that does not damage the system or component being inspected. X-ray is a system of nondestructive inspection that allows the inside of a device to be inspected without it having to be opened or in any way damaged.

(2) **Strength**. The ability of a material to resist distortion or deformation caused by an external force acting on it. Strength is the ability of a material to resist stresses that try to break it.

(3) **Neutron**. A particle in the nucleus of an atom which has mass, but no electrical charge. A neutron has about the same mass as a proton, and it adds weight to the atom, but does not affect its chemical behavior. Neutrons serve as a "cement" to hold the protons in the nucleus of an atom.

(4) **Delamination**. A separation of the layers of a laminated material. Plywood is made up of layers, or laminations, of wood glued together. When the glue weakens, the layers can separate and the plywood is said to be delaminated. Delamination of bonded composite materials causes the materials to lose strength.

(5) **Dye penetrant inspection**. A method of nondestructive inspection used to detect surface defects in metal or plastic parts. The part to be inspected is soaked in a penetrating liquid for a specified period of time, and then all the penetrant is

washed from its surface. The surface is then covered with a developing powder that pulls the penetrant from any defects that extend to the surface of the part. The penetrant plainly shows upon the powder-covered surface, outlining the defect.

(6) **Transducer**. A device that changes energy from one form into another. A microphone is an example of a transducer. It changes variations in air pressure into variations in electrical voltage. A radio speaker is another type of transducer. A speaker changes variation in electrical voltage into variations in air pressure, or sound.

(7) **Moisture**. Water contained in a mass of air. The water may be in the form of vapor, or it may be tiny droplets of haze, mist, or fog.

(8) **Galvanic corrosion**. A form of corrosion that occurs between two metals having different locations in the electrolytic, or galvanic, series. Corrosion can take place any time two metals are covered with an electrolyte. The amount of corrosion is determined by the difference in the location of the two metals in the galvanic series. The more active metal (the metal highest in the galvanic series) acts as the anode in the electrolytic process, and is eaten away. The more noble metal is not affected.

(9) **Intergranular corrosion**. A type of corrosion which forms along the boundaries of the grains inside a piece of metal alloy. Stress corrosion and exfoliation corrosion are types of inter-granular corrosion.

(10) **Stress corrosion**. A form of intergranular corrosion that forms in metals that are subject to a continuous tensile stress. The tensile stress separates the metal along the internal grain boundaries, and the corrosion acts at the apex of the cracks that form.

2. Answer the following questions in English.
(1) What are the methods of nondestructive testing?
(2) What is eddy current inspection?
(3) What is magnetic particle inspection?
(4) What is liquid dye penetrant inspection?
(5) What are the different types of materials corrosion?
(6) What are the characteristics of intergranular corrosion?
(7) Under what condition does galvanic corrosion occur?

SUPPLEMENTARY READING

[1] **Disposal and recycling of aerospace materials**: Materials selection must consider whole-of-life management issues, including the selection of sustainable materials that can be recycled using processes that are cheaper and more environmentally friendly than the

processes used to make new materials. The recycling of structural and engine materials is becoming more important to the aerospace industry with the increasing rate of aircraft retirements.

[2] The aerospace industry is moving towards high targets (above 80%) in the recycling of structural materials. About 60% of the airframe is currently recycled, but the industry is aiming to increase this to 80%. Careful consideration of the selection of sustainable materials in the design phase of aircraft is essential to ensure high levels of recycling.

[3] Recycling of metals is possible without any loss in mechanical performance, and these materials can be recycled and reused an infinite number of times without any detrimental effect on properties. The energy consumed in the recycling of metals is much less than the energy needed to extract metal from ore. The commercial incentive to recycle metals such as titanium, nickel, aluminum and magnesium is strong because of the high sale value of the scrap, whereas the value of steel is much less. The cost of removing steel components, cutting them into small pieces, transporting them to the recycling plant, and then recycling and casting the metal is greater than the sale value of the recycled steel. The main incentive for the aerospace industry to recycle steel is environmental rather than economic.

[4] The recycling of aluminum and steel components is performed using standard melting, refinement and casting processes. More specialist recycling processes are needed for the other metals, such as vacuum melting of titanium and nickel alloys to avoid excessive oxidation and to eliminate trapped gases in the molten metal.

[5] Recycling of fiber-polymer composites is difficult, particularly with thermoset matrix materials. Composites are recycled using grinding, incineration or chemical processes, although the cost of recycling is not competitive against the cost of new material. Furthermore, the fibers are weakened by grinding and thermal recovery processes, thus limiting their reuse in structural products requiring high strength.

TOPIC 5：简化技术英语

简化技术英语简介

ASD-STE100（Aerospace Defence Industries Association-Simplified Technical English 100，即简化技术英语）是一项国际性的航空航天标准，也是一种可控语言标准，有助于技术文档的简明性，让读者容易理解。此术语规范最初是为了飞机维修文档而设定的，其中包括飞机维护手册 aircraft maintenance manuals（AMM）、部件维护手册 component maintenance manuals（CMM）和服务通告 service bulletins（SB，由飞机制造商发出的文件），但现在也广泛地应用于汽车、机械、电子、电信和医疗设备等技术领域的技术文件。其内容包括：

(1) STE 的适用性；
(2) STE 和公司风格指南；
(3) STE 的写作规则；
(4) STE 核心概念：词汇（vocabulary）& 术语（terminology）；
(5) STE 带来的翻译获益；
(6) 在公司环境应用 STE；
(7) STE 应用实例和行业案例。

但需要注意的是，它并不适用于口语沟通。使用 STE 的优点包括：

(1) 减少歧义；
(2) 提高技术写作清晰度，尤其是过程步骤编写；
(3) 提高母语不是英语的人的理解力；
(4) 使人工翻译更容易、更快、更经济；
(5) 方便电脑辅助翻译和机器翻译；
(6) 通过降低维修和装配过程中出现错误的可能性来提高可靠性。

1. 简化技术英语的使用规则

ADS-STE100 的第一部分主要介绍了简化英语的使用规则，共有 14 条。通过对单词词性、含义的介绍，对照其使用的规则，帮助技术文件编写人员规范书写。

单词：

(1) You can use words that are：
- approved in the dictionary.
- technical names.
- technical verbs.

词性：

(2) Use approved words from the dictionary, only as part of speech given.

词义：

(3) Use approved words only with the approved meanings.

动词和形容词的形式：

（4）Use only the approved forms of verbs and adjectives.

技术名词：

（5）You can use words that you can include in a technical name category.

（6）Use the word that is unapproved in the dictionary only when it is a technical name or part of a technical name.

（7）Do not use words that are technical names as verbs.

（8）Use technique names that agree with the approved nomenclature.

（9）When you must select a technical name, use one which is a short and easy to understand.

（10）Do not use slang or jargon words as technical names.

（11）Do not use different technique names for the same item.

技术动词：

（12）You can use verbs that you can include in a Technical verb category.

（13）Do not use technique verbs as nouns.

拼写：

（14）Use American English spelling.

2. 可以使用的一般词汇

ADS-STE100 的第二部分是一个受控词典，给出了可以使用的一般词汇，每个词汇包含的信息有单词（词性）、已批准的标准/可替换的单词、批准示例、未经批准。

例 1：

Simplified Technical English, ASD-STE100

Word (part of speech)	Approved meaning/ ALTERNATIVES	APPROVED EXAMPLE	Not approved
TEST (n)	The procedure where an object or system is operated to make sure that its performance and/or function is correct	DISCONNECT ALL THE SYSTEMS WHICH ARE NOT NECESSARY FOR THE TEST.	
test (v)	TEST (n)	DO A FUNCTIONAL TEST OF THE WARNING SYSTEM.	Functionally test warning system.

test 是一个高频词，在 AMM 手册 501 为 adjustment/test（调整/测试）。在 ASD-STE100 中，对 test 的规范要求如下：

STE：Do the leak test of the system./Do a test for leaks in the system.

Non-STE：Test the system for leaks.

翻译：做系统泄漏测试。

"test"这个词在词典中是批准的名词，不是一个批准的动词。动词的"test"可以被名词的"test"替换。因此，在维护手册中，大多数"test"以名词形式出现。如：Do the test of the passenger reading light system.

例 2：

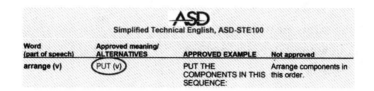

在 STE 中，arrange 是一个未经批准的单词，词典给出的替代词是"put"。
STE：Put the components in this sequence.
Non-STE：Arrange components in this order.
翻译：将组件按照顺序放置。

动词"put"在 STE 中是一个被批准的单词，意思是"放置"。除此之外，put 还是很多未被批准单词的替代词，大约有 26 个：arrange，bring，feed，guid，hook，inject，injection，insert，lay，lead，locate，mount，offer up，pack，repack，place，position，pour，prime，rest，route，station，submerge，thread，tuck，wrap。

ASD-STE100 的初衷就是简化技术英语，put 之所以被批准就是因为它简单易懂，其过去式和过去分词都是原形。日常维护手册中，它是一个高频词。通过"toolbox remote"系统对"put"单词查询，出现频次达 1.3 万次，可见该词出现频率之高。

例 3：

Play 这个单词出现频率较低，在手册中，它不能翻译为"玩耍、游玩"。在 ASD-ATE100 中，"play"是名词，翻译为"空间、间隙"。如：
Main landing gear beam hanger link free play check.
主起落架吊架连杆间隙检查。
If the free play is less than 0.050 in.
如果间隙小于 0.050 英寸。

例 4：

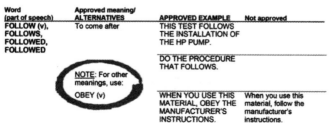

Follow 在 ASD-STE100 中,被批准的意思是跟随(v),这里特别注明:如果想表达其他含义,可以使用"obey 遵守(v)"。

STE:Obey the safety instructions.

Non-STE:Follow the safety instructions.

翻译:遵守安全须知。

3. 技术写作规则

(1) 短句的描述:一个句子只能有一个主题,不要为了使句子简短而省略词语或者结构。

(2) 表示步骤的句子长度:尽可能地保持句子的间断性,最长不要超过 20 个单词;每个句子只描述一个动作;仅仅在多个动作连续发生的情况下,才在一个句子中描述多个动作。

(3) 描述性写作:

① 句子长度最长不超过 25 个单词,尽量使用不同的句子长度和构造来保持稳定的可读性;

② 使用段落展示文档的逻辑结构,每个段落只有一个主题句;

③ 主题句通常位于段落开头的位置;

④ 段落最长不超过 6 个句子;

⑤ 每 10 个段落中,使用一句话的段落不要超过 1 次;

⑥ 对于新的、复杂的信息慢慢陈述。

(4) 警告、告诫、说明:

① 使用简单清晰的指令来开始一个**警告**或者**告诫**;

② **警告**信息和**告诫**事项要描述准确;

③ 如果可以,针对此**警告**或者**告诫**事项可能导致的风险,给出一个解决方案;

④ 仔细确认**警告**和**告诫**指令的正确性;

⑤ 如果某个条件是技术人员处理本条**警告**或者**告诫**的必要条件,把它放在最前面。

UNIT 6

ELECTRICAL SYSTEM

[1] Electrical systems have made significant advances over the years as aircraft has become more dependent upon electrically powered services. A typical electrical power system of the 1940s and 1950s was the twin 28 V DC system. This system was used a great deal on twin engined aircraft; each engine powered a 28 V DC generator which could employ load sharing with its contemporary if required. One or two DC batteries were also fitted and an inverter was provided to supply 115 V AC to the flight instruments.

[2] The advent of the V-bombers changed this situation radically due to the much greater power requirements-one the Vickers-Valiant-incorporated electrically actuated landing gear. They were fitted with four 115 V AC generators, one being driven by each engine. To provide the advantages of no-break power these generators were paralleled which increased the amount of control and protection circuitry. The V-bombers had to power high loads such as radar and electronic warfare jamming equipment. However, examination of the Nimrod maritime patrol aircraft (derived from the De-Havilland Comet) shows many similarities. As a yardstick of the rated power generated; the Victor (Fig. 6-1) was fitted with four 73 KVA AC generators while the Nimrod was fitted with four 60 KVA generators.

[3] Since that time, electrical systems had evolved in the manner shown in Fig. 6-2. In the UK, the introduction of powerful new AC electrical systems paved the way for the introduction of electrically powered power flying controls. Four-channel AC electrical systems utilised on the Avro Vulcan B2 and Handley Page Victor V-Bombers and the Vickers VC10 transport aircraft utilized flight control actuators powered by the aircraft AC electrical system rather than centralized aircraft hydraulic systems.

[4] Aircraft such as the Mc Donnell Douglas F-4 Phantom introduced high power AC generation systems to a fighter application. In order to generate constant frequency 115 V AC at 400 Hz a constant speed drive (CSD) is required to negate the aircraft engine speed variation over approximately 2 ∶ 1 speed range (full power speed; flight idle speed). These are complex hydromechanical devices which by their very nature are not highly reliable. Therefore, the introduction of constant frequency AC generation

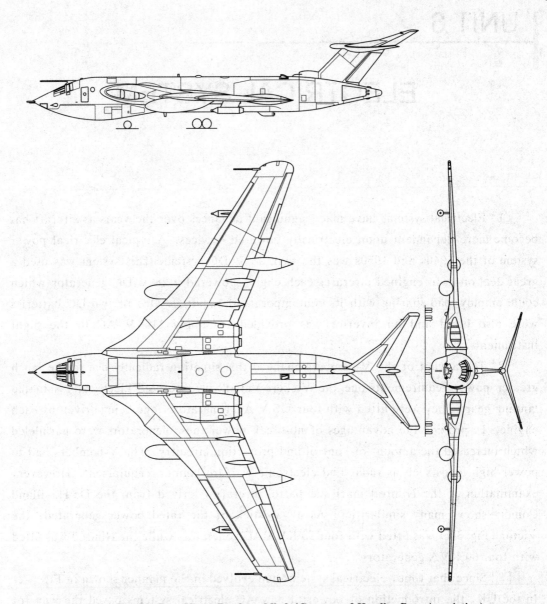

Fig. 6-1 Handley Page Victor Bomber, Mk 2 (Courtesy of Handley Page Association)

systems is not without accompanying reliability problems, particularly on fighter aircraft where engine throttle settings are changed very frequently throughout the mission.

[5] The advances in high power solid state switching technology together with enhancements in the necessary control electronics have made variable-speed/constant frequency (VSCF) systems a viable proposition in the last decade. The VSCF system remores the unreliable CSD portion; the variable frequency or frequency wild power from the AC generator being converted to 400 Hz constant frequency 115 V AC power by means of a solid state VSCF converter. VSCF systems are now becoming more commonplace: the F-18 fighter uses such a system and some versions of the Boeing 737-

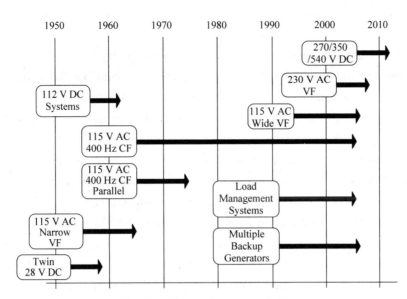

Fig. 6-2 Electrical system evolution

500 did use such a system, not with a lot of success in that particular case. In addition, the Boeing 777 airliner utilizes a VSCF system for backup AC power generation.

[6] In US military circles great emphasis is being placed by the US Air Force and the US Navy into the development of 270 V DC systems. In these systems high power generators derive 270 V DC power, some of which are then converted into 115 V AC 400 Hz or 28 V DC required to power specific equipments and loads. This approach has been adopted on the Lockheed Martin F-22 and F-35 aircraft. This is claimed to be more efficient than conventional methods of power generation and the amount of power conversion required is reduced with accompanying weight savings. These developments are allied to the "more-electric aircraft" concept where it is intended to ascribe more aircraft power system activities to electrical means rather than use hydraulic or high pressure bleed air which is presently the case. The fighter aircraft of tomorrow will therefore need to generate much higher levels of electrical power than at present. Schemes for the use of 270 V DC are envisaging power of 250 to 300 kW and possibly as much as 500 kW per channel; several times the typical level of 50 KVA per channel of today.

[7] At the component level, advances in the development of high power contactors and solid state power switching devices are improving the way in which aircraft primary and secondary power loads are switched and protected. These advances are being married to microelectronic developments to enable the implementation of new concepts for electrical power management system distribution, protection and load switching. The use of electrical power has progressed to the point where the generation, distribution and protection of electrical power to the aircraft electrical services or loads

now comprises one of the most complex aircraft systems. This situation was not always so.

[8] The move towards the higher AC voltage is really driven by the amount of power the electrical channel is required to produce. The sensible limit for DC systems has been found to be around 400 amps due to the limitations of feeder size and high power protection switchgear; known as contactors. Therefore, for a 28 V DC system delivering 400 amps, the maximum power the channel may deliver is about 12 kW, well below the requirements of most aircraft today. This level of power is sufficient for General Aviation (GA) aircraft and some of the smaller business jets. However, the requirements for aircraft power in business jets, regional aircraft and larger transport aircraft is usually in the range 20 to 90 KVA per channel and higher.

[9] The requirement for more power has been matched in the military aircraft arena. More recent aircraft have also readopted VF generation since this is the most reliable method at the generation level, though additional motor controllers may be needed elsewhere in the system to mitigate the effects of frequency variation. Power levels have increased steadily with the Airbus A380 utilising 150 KVA per channel and the Boeing 787 being even more electric with 500 KVA per channel. The status of More-Electric Aircraft (MEA) technologies and architectures are described in detail in the following paragraphs.

[10] For a number of years the concept of the "All Electric Aircraft" has been espoused. The Bristol Brabazon utilised a great number of electrical systems and the Vickers Valiant V-Bomber was also highly electrical in nature. At the time mid 1950s the concept did not fully catch on, though over the years there has been a great deal of debate relating to the advantages of electrical versus other forms of secondary power, such as hydraulics or high pressure bleed air systems.

[11] Over the past decade, examination of the benefits of the all-electric aircraft has been promoted by a number of aeronautical agencies in the US. In the early 1980s NASA funded a number of studies addressing the Integrated Digital Electrical Airplane (IDEA). The IDEA concept studies embraced a range of technologies which could improve the efficiency of a 250-300 seater replacement for an aircraft such as the Lockheed L1011 (Tristar).

[12] **Advanced electrical power systems**: The development of new systems to generate and distribute electrical power as an adjunct to more efficient engine power extraction flight control system and flight control actuation developments are already underway or are embodied in major civil programmes as evidenced by systems on the Airbus A380 and Boeing 787 aircraft. The A380 and B787 also use novel more-electric features as will be described elsewhere in this book.

[13] The generic parts of a typical Alternating Current aircraft electrical system are shown in Fig. 6-3 comprising the following:

- Power generation;
- Primary power distribution and protection;
- Power conversion and energy storage;
- Secondary power distribution and protection.

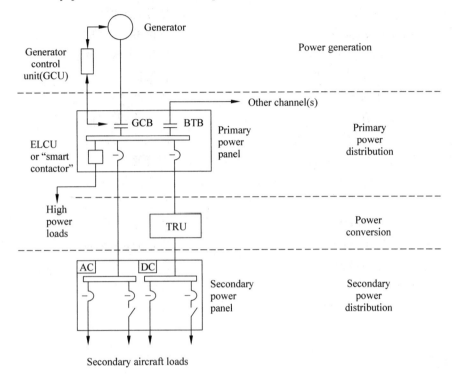

Fig. 6-3 Generic aircraft AC electrical system

NEW WORDS AND EXPRESSIONS

(1) electrical[ɪˈlektrɪkl]　　　　　电气
(2) electronic[ɪˌlekˈtrɒnɪk]　　　电子的,电子设备的
(3) load[ləʊd]　　　　　　　　　载荷,装载负荷
(4) generator[ˈdʒenəreɪtə(r)]　　　发电机
(5) DC: direct-current　　　　　　直流电
(6) AC: alternating current　　　　交流电
(7) KVA: kilovoltampere　　　　　千伏安
(8) incorporated[ɪnˈkɔːpəreɪtɪd]　采用,引入,并入的,联合的,使混合
(9) parallel[ˈpærəlel]　　　　　　平行的,并联
(10) protection[prəˈtekʃn]　　　　保护,预防
(11) circuitry[ˈsɜːkɪtrɪ]　　　　　电路,电路系统
(12) radar[ˈreɪdɑː(r)]　　　　　　雷达

(13) constant ['kɒnstənt]　　　　　常数,恒定的,持续的,不变的,常量
(14) frequency ['fri:kwənsɪ]　　　频率
(15) nature ['neɪtʃə(r)]　　　　　 自然,本性,固有特性
(16) throttle ['θrɒtl]　　　　　　 节流阀,油门
(17) mission ['mɪʃn]　　　　　　　任务,使命
(18) variable ['veərɪəbl]　　　　 可调的,可变的
(19) proposition [ˌprɒpə'zɪʃn]　　提议,主张;命题
(20) converter [kən'vɜːtə(r)]　　 转换器
(21) backup ['bækʌp]　　　　　　　备用的
(22) efficient [ɪ'fɪʃnt]　　　　　有效的,效率高的
(23) conventional [kən'venʃənl]　 常规的,传统的
(24) VSVF: variable-speed/variable frequency　变速变频
(25) CSCF: constant-speed/constant frequency　恒速恒频
(26) VSCF: variable-speed/constant frequency　变速恒频
(27) amplifier ['æmplɪfaɪə]　　　 放大器
(28) primary ['praɪmərɪ]　　　　　初始的,主要的
(29) secondary ['sekəndrɪ]　　　　次级的,次要的,辅助的
(30) distribution [ˌdɪstrɪ'bjuːʃn]　分配,分布
(31) conversion [kən'vɜːʃn]　　　 转换

EXERCISE

1. Translate the followings into Chinese.

(1) **Generator.**
   a. engine that converts mechanical energy into electrical energy by electromagnetic induction.
   b. an apparatus that produces a vapor or gas.
   c. an electronic device for producing a signal voltage.
   d. someone who originates or causes or initiates something.

(2) **CSD.** Constant Speed Drive. A hydro-mechanical component for changing variable engine accessory gearbox speed to constant speed to drive the electrical power.

(3) **AC.** Alternating Current. Electric current that changes direction is called AC or alternating current.

(4) **DC.** Direct Current. Electric current that always flows in the same direction.

(5) **Transformer.** A transformer is a static electrical device that transfers energy by inductive coupling between its winding circuits. A varying current in the primary winding creates a varying magnetic flux in the transformer's core and thus a varying magnetic flux through the secondary winding. This varying

magnetic flux induces a varying electromotive force (emf) or voltage in the secondary winding.

(6) **Rectifier**. A rectifier is an electrical device that converts alternating current (AC), which periodically reverses direction, to direct current (DC), which flows in only one direction. The process is known as rectification.

(7) **Primary power**. The primary electrical power system.

(8) **Secondary power**. Secondary electrical power source.

(9) **More-electric aircraft**. The More Electric Aircraft (MEA) concept is based on utilizing electric power to drive aircraft subsystems.

(10) **IDG (integrated drive generator)**. An AC generator installed on turbine engines. An IDG incorporates a brushless, three-phase AC generator and a constant-speed drive in a single component.

2. Answer the following questions in English.

(1) Please state the main functions of the electrical power.

(2) Please state the main constitutes of a typical electrical power system.

(3) Please state the main kinds of a typical electrical power system.

(4) What are the three kinds of the AC system?

(5) What does the MEA mean?

SUPPLEMENTARY READING

[1] The electrical power is used for the control, operation, and indication of the various airplane systems on the ground and inflight. The electrical power is obtained from the battery, generator and ground support equipment. It is controlled and monitored prior to distribution as 115 V AC, 28 V AC and 28 V DC supply to using systems.

[2] The AC generation supplies 115 V AC power to the airplane. The AC generation system has three subsystems: IDG power, APU generator power, and backup generator power. The IDGs are the primary sources of electrical power. The APU generator and backup generators are secondary sources of electrical power. Switches on the electrical panel control the IDG disconnect operations. GCUs control the IDGs and the APU generator. One backup generator converter controls the two backup generators. Light on the electrical panel and EICAS/ECAM messages show generator drive system information.

[3] Each engine mechanically turns an IDG and a backup generator. The IDG change the mechanical power to constant-frequency AC electrical power. The backup generators change the mechanical power to variable-frequency AC electrical power. The backup generator converter changes the variable-frequency AC power to constant frequency AC power. A continuous flow of fuel through the IDG fuel/oil heat exchanger

cools the IDG oil. A continuous flow of fan air through the backup generator/IDG air/oil heat exchanger cools the oil for the backup generator and IDG.

[4] The generators driven by the engine produce the AC power required by the airplane systems primarily for flight operation. The AC generator is rated at 50 KVA, 120/208 V, 400 Hz. The unit is without slip-rings commutator, or brushes on either the main generator or the exciter. A complete generator assembly consists of an exciter generator, a rotating rectifier and a main generator.

[5] APU generator is in the APU compartment. It attaches to the accessory pad of the APU generator. A seal plate is between the APU generator and the APU gearbox. The APU gearbox turns the APU generator. The APU GCU controls and monitors the operation of the APU generator. The APU generator supplies power through the auxiliary power breaker to the tie bus. The APU generator supplies power through the ground service select relay to the ground service bus. The APU generator also supplies power through the ground handling relay to the ground handling bus.

[6] The DC generator system changes AC power from the generators into DC power. It supplies power for DC loads and to charge the batteries. The batteries are an alternative power source for some loads when the generators do not operate. Major DC system components include main battery, battery charger, 2 transformer rectifier units (TRU) and static inverter located in the main equipment center and forward equipment area.

[7] Each TRU normally supplies its own DC load buses. An automatic operating DC tie control unit allows a single TRU to supply both main DC buses. During autoland, the centre DC bus receives power directly from the hot battery bus and center AC bus is supplied from the static inverter.

[8] The main battery/battery charger system provides DC power for the standby and autoland system. A separate APU battery/battery charger system supplies power for APU starting. In the charge mode battery charging is constant current and temperature compensated to prevent thermal runway. A thermistor and thermal switch, both integral to the battery, provide temperature sensing and overtemperature protection signals for charger control and shutdown. In the transformer rectifier unit mode, the battery charger supplies constant voltage. The battery current monitor is energized by the battery bus during normal flight operation. During autoland operation the battery current monitor is de-energized. When the main battery is discharging greater than 4 amperes, a ground signal is provided to illuminate the DISCH light. The signal also activates the "MAIN BAT DISCH" advisory level EICAS message.

[9] The battery is located in the electronic compartment, left side just forward of the E2 rack. The purpose of the battery is to provide DC power to critical airplane systems in absence of normal DC supply from the transformer rectifier. It is also used as a backup power for the AC system control and protection and for starting the APU.

Three transformer rectifier units are located in the electronic compartment on E3 rack. The purpose of the 3 transformer rectifiers is to convert 115 V AC, 400 Hz, 3-phase power to 28 V DC power for use by the airplane's systems.

[10] The AC and DC power is distributed to the various systems through the left and right load control centers in the flight compartment. The control and indication of the electrical system is from the P5 overhead panel in the flight compartment. The purpose of electrical power distribution system is to provide and control generated AC and DC power for use by the various airplane systems, the distribution system consists of 115 V AC, 28 V AC and 28 V DC power obtained from generators and battery.

[11] The external power system controls the electrical power on the airplane from ground power sources. The components of the external power system are as follows:
- External power contactor;
- Ground service transfer relay;
- Ground service select relay;
- Ground handling relay;
- BPCU;
- Ground handling transformer rectifier unit.

The BPCU (Bus power control unit) controls external power contactor, also controls ground handling and ground service power. Ground handling power is available only on the ground. Ground service power is available on the ground and in flight.

[12] A ground handling bus powers equipment used during ground operation. The bus is not connected to the airplane main AC buses and is supplied from external power or the APU generator only. When the good power (power quality must be satisfactory) apply to the external power receptacle of airplane, the BPCU energizes one coil (external power coil) of the ground handling relay. This connect the external power to the ground handling bus, if no external power is available, but the APU is running, the BPCU energizes the other coil (APU power coil) of the ground handling relay. This connects the APU generator to the ground handling bus.

[13] The ground service bus supplies power to inflight loads and can provide power on the ground for airplane cleaning and servicing operations. The ground service bus is energized from external power, APU generator or the right main bus. When the good power applies to the external power receptacle of airplane, power is available at the ground service transfer relay. When the ground service switch is pushed, the BPCU energizes the ground service transfer relay. This connects the external power to the ground service bus. When the ground service switch is pushed with no external power available, but the APU running, the BPCU energizes the ground service select relay. This connects the APU generator to the ground service bus. When the right main AC bus has power, it automatically supplies the power to the ground service bus. In this case, the ground service switch has no effect on the ground service bus.

TOPIC 6：简化技术英语（续）

技术动词

技术动词是在特定的技术和操作环境中给出指示和信息的词。字典中不包括技术动词，因为技术动词太多，每个制造商使用不同的词来描述相同的动作。

简化技术英语（STE）提供了一个类别列表，并附有示例，以帮助正确使用技术动词。如果单词属于以下四个类别之一，则它们是技术动词。

1) Manufacturing processes 制造工艺

（1）Remove material 去除材料：

drill 钻，grind 磨，mill 磨，ream 铰

（2）Add material 添加材质：

flame 燃烧，insulate 绝缘，remetal 重熔，retread 翻新

（3）Attach material 附加材料：

bond 黏结，braze 钎焊，crimp 压接，rivet 铆接，solder 焊接，weld 焊接

（4）Change the mechanical strength, the structure, or physical properties of a material 改变材料的机械强度、结构或物理特性：

anneal 退火，cure 固化，freeze 冷冻，heat-treat 热处理，magnetize 磁化，normalize 正火

（5）Change the surface finish of a material 更改材质的表面光洁度：

buff 擦光，burnish 抛光，passivate 钝化，plate 电镀，polish 抛光

（6）Change the shape of a material 更改材质的形状：

cast 铸造，extrude 挤压，spin 旋转，stamp 压印

2) Computer processes and applications 计算机处理和应用

（1）Input/output processes 输入/输出处理：

enter 输入，click 单击，digitize 数字化，print 打印

（2）User interface and application processes 用户界面和应用程序处理：

clear 清除，close 关闭，copy 复制，cut 剪切，delete 删除，drag 拖动，drag-and-drop 拖放，encrypt 加密，erase 擦除，filter 筛选，highlight 突出显示，maximize 最大化，minimize 最小化，open 打开，paste 粘贴，save 保存，scroll 滚动，sort 排序，store 存储

（3）System operations 系统操作：

abort 中止，boot 启动，communicate 通信，debug 调试，download 下载，format 格式化，install 安装，load 加载，manage 管理，process 处理，reboot 重新启动，update 更新，upgrade 升级，upload 上传

3) Descriptions 描述

这类技术动词仅适用于描述性文本，如通用信息、系统描述和操作文本以及服务公告的描述部分。在编写过程时不要使用以下动词：

（1）Mathematical，scientific and engineering processes 数学、科学和工程处理：

bisect 平分，compensate for 补偿，detect 检测，emit 发射，modulate 调制，radiate 发射

（2）Military processes 军事处理：

aim 瞄准，arm 武装，detect 探测，disable 禁用，enable 启用，explode 爆炸，fire 开火，intercept 拦截，load 装载，parachute 伞投，unload 卸载

（3）Regulatory language 监管语言：

waive（for inspection and requirements）放弃（检查和要求），comply with 遵守，conform to 遵守，supersede 取代，meet（a requirement）满足（要求）

4）Operational language 操作语言

这类技术动词只适用于涉及操作上下文的文本，操作性文本告诉用户如何正确操作和使用某物。例如，有关如何使用电话、平板电脑、医疗设备或电视机的手册是一个操作文本。示例：

airdrop 空投	alert 警戒	approach 进近
authorize 授权	brief 简要	evacuate 疏散
contact 联络	crank 冷转	descend 下降
deviate 偏航	disembark 登陆	drift 漂移
dry-motor 干冷转	enable 启用	call 呼叫
fasten 系紧	ferry 摆渡	hover 悬停
inform 通知	inhibit 禁止	land 着陆
load 装载	maintain 维护	navigate 导航
observe 观察	provide 提供	reach 到达
respond 响应	retard 减速	retrim 再入
return 返回	rotate 旋转	serve 发球
shut down 关机	sideslip 侧滑	sit 坐下
sleep 睡觉	switch off 关机	trim 配平
take off 起飞	take over 接管	taxi 滑行
tie 系紧	trigger 触发	switch on 开机
unfasten 松开	unlatch 解锁	unload 卸载
verify 验证	wet-motor 湿冷转	

以上所列的技术动词只是例子，并不是所有技术动词的完整列表。

简化技术英语示例：

（1）Non-STE：If you **detect** broken wires, repair them.

　　非 STE：如果检测到电线断了，修好它们。

　　STE：If you **find** broken wires, repair them.

　　STE：如果发现电线断了，修好它们。

（2）Non-STE：**Machine** the hole until it has a diameter of （8.00±0.003）mm.

　　非 STE：机加工孔，直到其直径为（8.00±0.003）mm。

　　STE：**Ream** the hole until it has a diameter of （8.00±0.003）mm.

　　STE：扩孔，直到其直径为（8.00±0.003）mm。

(3) Non-STE：**Grease** the fasteners.

非 STE：润滑紧固件。

STE：**Apply** grease to the fasteners.

STE：在紧固件上涂抹润滑脂。

(4) Non-STE：**Wire** the cable to the structure.

非 STE：将钢索连接到结构上。

STE：**Attach** the cable to the structure with wire.

STE：用金属线把钢索连接到结构上。

(5) Non-STE：Do not **enter** the engine test area without approval.

非 STE：未经批准，不得进入发动机试车区域。

STE：Do not **go into** the engine test area without approval.

STE：未经批准，请勿进入发动机试车区域。

(6) Non-STE：The door is made of carbon **fibre** reinforced plastic.

非 STE：门由碳纤维增强塑料制成。("fibre"是英国英语拼写)

STE：The door is made of carbon **fiber** reinforced plastic.

STE：门是用碳纤维增强塑料做的。("fiber"是美国英语拼写)

(7) Non-STE：Change the **colour** of the display.

非 STE：改变显示器的颜色。("colour"是英国英语拼写)

STE：Change the **color** of the display.

STE：改变显示器的颜色。("color"是美式英语拼写)

UNIT 7

AVIONICS SYSTEM

[1] In order to safely fly any aircraft, a pilot must understand how to interpret and operate the flight instruments. The pilot also needs to be able to recognize associated errors and malfunctions of these instruments. This section addresses the pitot-static system and associated instruments, the vacuum system and related instruments, gyroscopic instruments, and the magnetic compass. When a pilot understands how each instrument works and recognizes when an instrument is malfunctioning, he or she can safely utilize the instruments to their fullest potential.

[2] The pitot-static system is a combined system that utilizes the static air pressure, and the dynamic pressure due to the motion of the aircraft through the air. These combined pressures are utilized for the operation of the airspeed indicator (ASI), altimeter, and vertical speed indicator (VSI), as shown in Fig. 7-1.

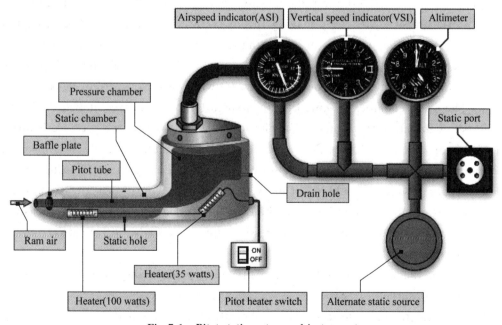

Fig. 7-1 Pitot-static system and instruments

[3] The pitot tube is utilized to measure the total combined pressures that are present when an aircraft moves through the air. Static pressure, also known as ambient

pressure, is always present whether an aircraft is moving or at rest. It is simply the barometric pressure in the local area. Dynamic pressure is present only when an aircraft is in motion; therefore, it can be thought of as a pressure due to motion. Wind also generates dynamic pressure. It does not matter if the aircraft is moving through still air at 70 knots or if the aircraft is facing a wind with a speed of 70 knots, the same dynamic pressure is generated.

[4] When the wind blows from an angle less than 90° off the nose of the aircraft, dynamic pressure can be depicted on the ASI. The wind moving across the airfoil at 20 knots is the same as the aircraft moving through calm air at 20 knots. The pitot tube captures the dynamic pressure, as well as the static pressure that is always present.

[5] The pitot tube has a small opening at the front which allows the total pressure to enter the pressure chamber. The total pressure is made up of dynamic pressure plus static pressure. In addition to the larger hole in the front of the pitot tube, there is a small hole in the back of the chamber which allows moisture to drain from the system should the aircraft enter precipitation. Both openings in the pitot tube need to be checked prior to flight to ensure that neither is blocked. Many aircraft have pitot tube covers installed when they sit for extended periods of time. This helps to keep bugs and other objects from becoming lodged in the opening of the pitot tube.

[6] The one instrument that utilizes the pitot tube is the ASI. The total pressure is transmitted to the ASI from the pitot tube's pressure chamber via a small tube. The static pressure is also delivered to the opposite side of the ASI which serves to cancel out the two static pressures, thereby leaving the dynamic pressure to be indicated on the instrument. When the dynamic pressure changes, the ASI shows either increase or decrease, corresponding to the direction of change. The two remaining instruments (altimeter and VSI) utilize only the static pressure which is derived from the static port.

[7] The static chamber is vented through small holes to the free undisturbed air on the side(s) of the aircraft. As the atmospheric pressure changes, the pressure is able to move freely in and out of the instruments through the small lines which connect the instruments into the static system. An alternate static source is provided in some aircraft to provide static pressure should the primary static source become blocked. The alternate static source is normally found inside of the flight deck. Due to the venturi effect of the air flowing around the fuselage, the air pressure inside the flight deck is lower than the exterior pressure.

[8] When the alternate static source pressure is used, the following instrument indications are observed:
- The altimeter indicates a slightly higher altitude than actual;
- The ASI indicates an airspeed greater than the actual airspeed;
- The VSI shows a momentary climb and then stabilizes if the altitude is held constant.

[9] Each pilot is responsible for consulting the aircraft flight manual (AFM) or the pilot's operating handbook (POH) to determine the amount of error that is introduced into the system when utilizing the alternate static source. In an aircraft not equipped with an alternate static source, an alternate method of introducing static pressure into the system should a blockage occur is to break the glass face of the VSI. This most likely renders the VSI inoperative. The reason for choosing the VSI as the instrument to break is that it is the least important static source instrument for flight.

[10] Course interceptions are performed in most phases of instrument navigation. The equipment used varies, but an intercept heading must be flown that results in an angle or rate of intercept sufficient to solve a particular problem.

[11] Rate of intercept, seen by the aviator as bearing pointer or horizontal situation indicator(HSI) movement, is a result of the following factors:
- The angle at which the aircraft is flown toward a desired course (angle of intercept);
- True airspeed and wind (ground speed, GS);
- Distance from the station.

[12] The angle of intercept is the angle between the heading of the aircraft (intercept heading) and desired course. Controlling this angle by selection/adjustment of the intercept heading is the easiest and most effective way to control course interceptions. Angle of intercept must be greater than the degrees from course, but should not exceed 90°. Within this limit, adjust to achieve the most desirable rate of intercept.

[13] When selecting an intercept heading, the key factor is the relationship between distance from the station and degrees from the course. Each degree, or radial, is 1 NM (nautical mile) wide at a distance of 60 NM from the station. Width increases or decreases in proportion to the 60 NM distance. For example, 1 degree is 2 NM wide at 120 NM and $\frac{1}{2}$ NM wide at 30 NM. For a given GS and angle of intercept, the resultant rate of intercept varies according to the distance from the station:
- Degrees from course;
- Distance from the station;
- True airspeed and wind (GS).

[14] Distance measuring equipment (DME) consists of an ultrahigh frequency (UHF) navigational aid with VOR (VHF omnidirectional radio range)/DMEs and VORTACs (visual omni-range tactical air navigation). It measures, in NM, the slant range distance of an aircraft from a VOR/DME or VORTAC (both hereafters referred to as a VORTAC). Although DME equipment is very popular, not all aircraft are DME equipped.

[15] To utilize DME, the pilot should select, tune, and identify a VORTAC, as previously described. The DME receiver, utilizing what is called a "paired frequency"

concept, automatically selects and tunes the UHF DME frequency associated with the VHF VORTAC frequency selected by the pilot. This process is entirely transparent to the pilot. After a brief pause, the DME display shows the slant range distance to or from the VORTAC. Slant range distance is the direct distance between the aircraft and the VORTAC, and is therefore affected by aircraft altitude. (Station passage directly over a VORTAC from an altitude of 6,076 feet above ground level (AGL) would show approximately 1.0 NM on the DME.) DME is a very useful adjunct to VOR navigation. A VOR radial alone merely gives line of position information. With DME, a pilot may precisely locate the aircraft on that line (radial).

[16] Most DME receivers also provide GS and time-to-station modes of operation. The GS is displayed in knots. The time-to-station mode displays the minutes remaining to VORTAC station passage, predicated upon the present GS. GS and time-to-station information is only accurate when tracking directly to or from a VORTAC/DME receivers typically need a minute or two of stabilized flight directly to or from a VORTAC before displaying accurate GS or time to-station information.

[17] Some DME installations have a hold feature that permits a DME signal to be retained from one VORTAC while the course indicator displays course deviation information from an ILS or another VORTAC.

[18] Area navigation (RNAV) permits electronic course guidance on any direct route between points established by the pilot. While RNAV is a generic term that applies to a variety of navigational aids, such as LORAN (long range aid to air navigation), GPS (global positioning system), and others. This section deals with VOR/DME-based RNAV. VOR/DME RNAV is not a separate ground-based NAVAID (navigation aid), but a method of navigation using VOR/DME and VORTAC signals specially processed by the aircraft's RNAV computer (Fig. 7-2).

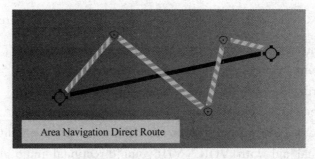

Fig. 7-2　Flying an RNAV course

[19] In its simplest form, VOR/DME RNAV allows the pilot to electronically move VORTACs around to more convenient locations. Once electronically relocated, they are referred to as waypoints. These waypoints are described as a combination of a selected radial and distance within the service volume of the VORTAC to be used. These waypoints allow a straight course to be flown between almost any origin and destination,

without regard to the orientation of VORTACs or the existence of airways.

[20] While the capabilities and methods of operation of VOR/DME RNAV units differ, there are basic principles of operation that are common to all. Pilots are urged to study the manufacturer's operating guide and receive instruction prior to the use of VOR/DME RNAV or any unfamiliar navigational system. Operational information and limitations should also be sought from placards and the supplement section of the AFM/POH (aircraft flight manual/pilot's operating handbook).

[21] VOR/DME-based RNAV units operate in at least three modes: VOR, enroute, and approach. A fourth mode, VOR Parallel, may also be found on some models. The units need both VOR and DME signals to operate in any RNAV mode. If the NAVAID selected is a VOR without DME, RNAV mode will not function.

[22] In the VOR (or non-RNAV) mode, the unit simply functions as a VOR receiver with DME capability (Fig. 7-3). The unit's display on the VOR indicator is conventional in all respects. For operation on established airways or any other ordinary VOR navigation, the VOR mode is used.

Fig. 7-3 RNAV controls

[23] To utilize the unit's RNAV capability, the pilot selects and establishes a waypoint or a series of waypoints to define a course. To operate in any RNAV mode, the unit needs both radial and distance signals; therefore, a VORTAC (or VOR/DME) needs to be selected as a NAVAID. To establish a waypoint, a point somewhere within the service range of a VORTAC is defined on the basis of radial and distance. Once the waypoint is entered into the unit and the RNAV en route mode is selected, the CDI (course deviation indicator) displays course guidance to the waypoint, not the original VORTAC. DME also displays distance to the waypoint. Many units have the capability to store several waypoints, allowing them to be programmed prior to flight, if desired, and called up in flight.

[24] RNAV waypoints are entered into the unit in magnetic bearings (radials) of degrees and tenths (i.e., 275.5°) and distances in NM and tenths (i.e., 25.2 NM). When plotting RNAV waypoints on an aeronautical chart, pilots find it difficult to measure to that level of accuracy, and in practical application, it is rarely necessary. A

number of flight planning publications publish airport coordinates and waypoints with this precision and the unit accepts those figures. There is a subtle, but important difference in CDI operation and display in the RNAV modes.

[25] In the RNAV modes, course deviation is displayed in terms of linear deviation. In the RNAV en route mode, maximum deflection of the CDI typically represents 5 NM on either side of the selected course, without regard to distance from the waypoint. In the RNAV approach mode, maximum deflection of the CDI typically represents $1\frac{1}{4}$ NM on either side of the selected course. There is no increase in CDI sensitivity as the aircraft approaches a waypoint in RNAV mode.

[26] The RNAV approach mode is used for instrument approaches. Its narrow scale width ($\frac{1}{4}$ of the enroute mode) permits very precise tracking to or from the selected waypoint. In visual flight rules (VFR) cross-country navigation, tracking a course in the approach mode is not desirable because it requires a great deal of attention and soon becomes tedious.

[27] A fourth, lesser-used mode on some units is the VOR parallel mode. This permits the CDI to display linear (not angular) deviation as the aircraft tracks to and from VORTACs. It derives its name from permitting the pilot to offset (or parallel) a selected course or airway at a fixed distance of the pilot's choosing, if desired. The VOR parallel mode has the same effect as placing a waypoint directly over an existing VORTAC. Some pilots select the VOR parallel mode when utilizing the navigation (NAV) tracking function of their autopilot for smoother course following near the VORTAC.

[28] Confusion is possible when navigating an aircraft with VOR/DME-based RNAV, and it is essential that the pilot become familiar with the equipment installed. It is not unknown for pilots to operate inadvertently in one of the RNAV modes when the operation was not intended by overlooking switch positions or annunciators. The reverse has also occurred with a pilot neglecting to place the unit into one of the RNAV modes by overlooking switch positions or annunciators. As always, the prudent pilot is not only familiar with the equipment used, but never places complete reliance in just one method of navigation when others are available for cross-check.

NEW WORDS AND EXPRESSIONS

(1) avionics[ˌeɪvɪˈɒnɪks]　　　　　　　　　航空电子，航空电子设备
(2) instrument[ˈɪnstrəmənt]　　　　　　　仪表
(3) associate[əˈsəʊʃɪeɪt]　　　　　　　　　有联系的
(4) malfunction[ˌmælˈfʌŋkʃn]　　　　　　故障，失效

(5) pitot['pɪtəʊ] 皮托管
(6) vacuum['vækjʊəm] 真空
(7) gyroscopic[ˌdʒaɪrə'skɒpɪk] 陀螺仪的
(8) magnetic[mæɡ'netɪk] 磁性的,磁的
(9) compass['kʌmpəs] 罗盘,指南针
(10) utilize['juːtəlaɪz] 利用
(11) potential[pə'tenʃl] 潜在的,电势
(12) static['stætɪk] 静止的,静力学的
(13) dynamic[daɪ'næmɪk] 动力学的
(14) ASI:air speed indicator 空速表
(15) VSI:vertical speed indicator 垂直速度表
(16) knot[nɒt] 节(速度单位)
(17) plus[plʌs] 加
(18) drain[dreɪn] 排水,放油
(19) precipitation[prɪˌsɪpɪ'teɪʃn] 降水,沉淀
(20) block[blɒk] 堵塞
(21) increase[ɪn'kriːs] 增加
(22) decrease[dɪ'kriːs] 减少
(23) corresponding[ˌkɒrə'spɒndɪŋ] 相应的,符合的
(24) derived[dɪ'raɪvd] 起源,由来,分支电路
(25) alternate[ɔːl'tɜːnət] 交替,备降的,备用的,备降机场
(26) AFM:aircraft flight manual 飞机飞行手册
(27) POH:pilot's operating handbook 飞行员操作手册
(28) intercept[ˌɪntə'sept] 拦截
(29) GS:ground speed 地速
(30) NM:nautical mile 海里
(31) DME:distance measuring equipment 测距仪
(32) VORTAC:visual omni-range tactical air navigation 目视全向战术空中导航
(33) RNAV:area navigation 区域导航
(34) LORAN:long range aid to air navigation 远距空中导航设备
(35) NAVAID:navigation aid 助航设备
(36) CDI:course deviation indicator 偏航指示器
(37) VFR:visual flight rules 目视飞行规则

EXERCISE

1. Translate the followings into Chinese.

(1) **Avionics**. Science and technology of electronic systems and devices for

aeronautics and astronautics. e. g. , avionics has become even more important with the development of the space program.

(2) **ASI**. Diagram showing the face of a true airspeed indicator typical for a faster single engine aircraft.

(3) **VSI**. An instrument that gives the pilot an indication of the rate at which an aircraft is rising or descending. A vertical-speed indicator is a pressure instrument with a calibrated restriction between the case and a pressure bellows. When the aircraft changes altitude, the pressure inside the case changes immediately. But because of the calibrated restriction, the pressure change inside the bellows lags behind the pressure change in the case.

(4) **VFR**. Flight rules adopted by the Federal Aviation Administration to govern aircraft flight when the pilot has visual reference to the ground at all times. VFR operations specify the amount of ceiling (the distance between the surface of the earth and the base of the clouds) and the visibility (the horizontal distance the pilot can see) that the pilot must have in order to operate according to these rules. When the weather conditions are not such that the pilot can operate according to VFR, he or she must use another set of rules, instrument flight rules (IFR).

(5) **CDI**. An indicator used with VOR, LORAN, or GPS navigation systems to indicate the direction and relative amount of deviation from the desired course. CDIs are often called Left-Right indicators.

(6) **Navigation**. The guidance of ships or airplanes from place to place.

(7) **GS**. The speed of a flying aircraft measured in relation to the ground it is traveling over and used for calculating flight times.

(8) **Distance measuring equipment (DME)**.

    a. Instrumentation that may be used to measure distances or lengths of objects, such as fiber optic displacement sensors that use fiber optic interferometric sensors.

    b. In radio location systems, equipment that ascertains the distance between an interrogator and a transponder by (a) measuring the delay of a signal to and from the transponder, (b) applying a known signal propagation velocity, (c) calculating the roundtrip distance.

    c. A radio navigation aid in the aeronautical radio navigation service that determines the distance of an interrogator from a transponder by measuring the propagation delay, i.e., the transit time, of a pulse to and from the transponder.

(9) **UHF**. Frequencies of electromagnetic radiation between 300 and 3,000 megahertz. The wavelengths for ultrahigh frequencies range between 100 and 10 centimeters. The UHF band is between the VHF (very-high-frequency) and the SHF (super-high-frequency) bands.

(10) **VOR**. A type of electronic navigation equipment in which the instrument in the

cockpit identifies the radial, or line from the VOR station measured in degrees clockwise from magnetic north, along which the aircraft is located. VOR is a phase-comparison system in which two signals transmitted simultaneously from a ground station are in phase only when they are received at a location directly magnetic north of the station. At a location magnetic east of the station, they are 90° out of phase. At magnetic south, they are 180° out of phase, and west of the station, they are 270° out of phase.

2. Answer the following questions in English.
(1) What are the main functions of the pitot-static system?
(2) What is the pitot-static system utilized for?
(3) What does the AFM mean?
(4) What does the POH mean?
(5) What does the DME consist of?

SUPPLEMENTARY READING

[1] The static and total air pressure system gets air pressure inputs from three pitot probes and six static ports on the airplane fuselage. These are the two types of air pressure:
- Static air pressure is the ambient air pressure around the airplane;
- Pitot air pressure is the air pressure on the pitot probe tube as a result of the forward motion of the airplane.

[2] The static and total air pressure system has these components:
- Three pitot probes;
- Six static ports;
- Five drain fittings.

[3] Flexible and hard pneumatic tubing are used to connect the pitot-static components. The system drains are used to remove condensation in the pitot-static lines.

[4] Primary Static and Total Air Pressure System: the two primary pitot probes connect to two pitot air data modules (ADMs). Two pairs of the primary static ports connect to two static ADMs. The ADMs change the air pressures to electrical signals and send them to the air data inertial reference units (ADIRUs) on ARINC (aeronautical radio incorporated) 429 data buses. The ADIRUs use the signals to calculate flight parameters such as airspeed and altitude. Each pitot line and each static line has a drain fitting.

[5] The auxiliary pitot probe connects to the standby altimeter/airspeed indicator. The alternate static ports connect to the standby altimeter/airspeed indicator and to the cabin differential pressure indicator.

[6] The standby static line has a drain fitting. The drain fittings for the captain and

first officer pitot lines are in the forward equipment compartment adjacent to the compartment access door. The pitot probe port points forward to measure pitot pressure. A strut moves the probe several inches from the airplane skin to decrease airflow turbulence effects. A base plate contains the electrical and pressure connectors. A gasket is between the probe base and the airplane structure to form a pressure seal.

[7] An anti-icing heater is in the probe to prevent ice. The heater is attached to the electrical connector in the base plate.

[8] The system drains remove condensation that collects in the pitot-static lines. The drain fittings have a reinforced transparent section of tubing with an orange float. This forms a sight gage to show the level of fluid that collects in the sump.

[9] The lower part of the drain contains a valve covered by a bayonet cap. To drain the pitot-static line, remove the cap and insert the valve depressor on the cap into the valve. The liquid in the sump drains by gravity when you push the valve.

[10] The standby altimeter/airspeed indicator is two flight instruments in one component. One instrument is a pneumatic altimeter. It gets static air pressure from the alternate static ports and shows barometric altitude.

[11] The other instrument is a pneumatic airspeed indicator. This indicator gets pitot air pressure from the alternate pitot probe and static air pressure from the alternate static ports to show the indicated air speed.

[12] The standby altimeter/airspeed indicator shows air data information such as barometric altitude and indicated airspeed. The standby altimeter/airspeed indicator receives pitot pressure from the alternate pitot probe. It also receives static pressure from the alternate static ports.

[13] The digital altitude display shows barometric altitude from 1,000 to 50,000 feet in thousands of feet. The altitude dial and pointer show altitude in 20-foot increments. One turn of the pointer is equal to 1,000 feet.

[14] Barometric pressure reference in inches of mercury (in Hg) and millibars (mbar) shows on the indicator. Rotate the barometric reference control knob to set the barometric pressure reference. It is adjustable from 745 to 1,049.5 mbar and from 22.00 to 30.99 inHg.

[15] The zero adjust screw lets you adjust the altitude without changing the barometric reference settings. The digital airspeed display shows airspeeds from 60 to 450 knots. A drum, graduated in knots, shows the indicated airspeed.

[16] A vibrator is on the instrument frame to reduce friction errors in the mechanical linkage and to improve indicator response. The vibrator receives 28 V DC power from the 28 V DC standby bus.

[17] The standby altimeter gets static pressure from the two alternate static ports.

The pressure sensor moves in response to the static pressure change. The sensor drives the altitude gear train which connects mechanically to the altitude counter and pointer.

[18] Rotating the "BARO" set knob on the front panel corrects for local changes in barometric pressure. The Standby Airspeed Indicator mechanism has two pressure ports. One port is for static pressure and the other port is for pitot pressure. The pressure sensor expands and contracts with changes in the pitot-static pressure. The movement is transmitted to a counter to indicate airspeed.

TOPIC 7：常用数学符号和公式读法

在科技英语中还经常会碰到一些高等数学、线性代数、积分变换、复变函数等多门数学学科的表达式或公式，下面介绍一些常用数学符号和公式的读法：

1st	first
2nd	second
3rd	third
4th	fourth
5th	fifth
9th	ninth
12th	twelfth
21st	twenty-first
1/2	one half
2/3	two thirds
19/56	nineteen over fifty-six
79.3	seventy-nine point three
3^2	three squared
5^3	five cubed
6^{10}	six to the power of ten
$\sqrt{x}$	square root of x
+	plus；positive
−	minus；negative
×	multiplied by；times
÷；/	divided by；over
=	is equal to；equals
±	plus or minus
<	less than
>	greater than
≫	much greater than
(	open/left parenthesis
)	close/right parenthesis
[	open/left bracket
]	close/right bracket
{	open/left brace
}	close/right brace
$\|a\|$	absolute value of a
a'	a prime

b''	b double prime
a_n	a sub n
d/dx	the derivative with respect to x; d by dx
$\partial/\partial x$	the partial derivative with respect to x
$\log_b a$	log to the base b of a
$\ln x$	the natural log of x
$p(x)$	p of x
Σ	summation
$\lim\limits_{x \to a} y$	limit as x approaches a of y
$\int f(x)dx$	integral of f of dx
$x \to \infty$	x approaches to infinity

UNIT 8

AIR CONDITIONING SYSTEM

[1] The function of an air conditioning system is to maintain a comfortable air temperature within the aircraft fuselage. The system will increase or decrease the temperature of the air as needed to obtain the desired value. Most systems are capable of producing an air temperature of 70-80°F with normally anticipated outside air temperatures. This temperature conditioned air is then distributed so that there is a minimum of stratification (hot and cold layers). The system, in addition, must provide the control of humidity, it must prevent the fogging of windows, and it must maintain the temperature of wall panels and floors at a comfortable level.

[2] An air conditioning system is designed to perform any or all of the following functions: (1) Supply ventilation air, (2) supply heated air, (3) supply pressurization air, and (4) supply cooling air. There are two types of air conditioning systems commonly used on aircraft. Air cycle air conditioning is used on most turbine-powered aircraft. It makes use of engine bleed air or APU pneumatic air during the conditioning process. Vapor cycle air conditioning systems are often used on reciprocating aircraft. This type of system is similar to that found in homes and automobiles. Note that some turbine-powered aircraft also use vapor cycle air conditioning.

[3] **Air cycle cooling system:** An air cycle cooling system consists of an expansion turbine (cooling turbine), an air-to-air heat exchanger, and various valves which control airflow through the system. High pressure air from the compressor is routed through the turbine section. As the air passes through the turbine, it rotates the turbine and the compressor. When the compressed air performs the work of turning the turbine, it undergoes a pressure and temperature drop. It is this temperature drop which produces the cold air used for air conditioning. Before entering the expansion turbine, the pressurized air is directed through an air-to-air heat exchanger. This unit utilizes outside air at ambient temperature to cool the compressed air. It should be evident that the heat exchanger can only cool the compressed air to the temperature of the ambient air temperature. The primary purpose of the heat exchanger is to remove the heat of compression so that the expansion turbine receives relatively cool air on which to start its own cooling process. Fig. 8-1 diagrams the air cycle cooling system of the B737-700.

UNIT 8 AIR CONDITIONING SYSTEM

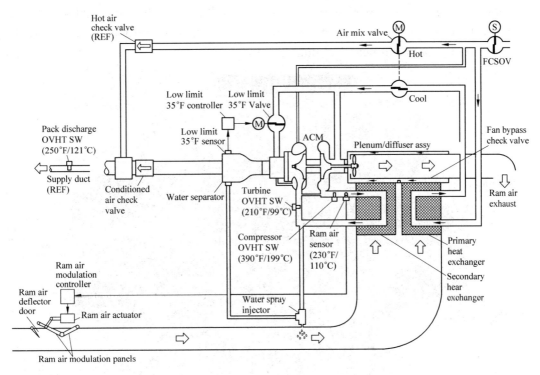

Fig. 8-1 Air cycle cooling system

[4] **Pack valve:** The pack valve is the valve that regulates bleed air from the pneumatic manifold into the air cycle air conditioning system. It is controlled with a switch from the air conditioning panel in the cockpit. Many pack valves are electrically controlled and pneumatically operated. Also known as the supply shutoff valve, the pack valve opens, closes, and modulates to allow the air cycle air cooling system to be supplied with a designed volume of hot, pressurized air. When an overheat or other abnormal condition requires that the air conditioning pack be shut down, a signal is sent to the pack valve to close.

[5] **Air cycle machine:** The heart of the air cycle cooling system is the refrigeration turbine unit, also known as the air cycle machine (ACM). It is comprised of a compressor that is driven by a turbine on a common shaft. System air flows from the primary heat exchanger into the compressor side of the ACM. As the air is compressed, its temperature rises. It is then sent to a secondary heat exchanger, similar to the primary heat exchanger located in the ram air duct. The elevated temperature of the ACM compressed air facilitates an easy exchange of heat energy to the ram air.

[6] The cooled system air, still under pressure from the continuous system air flow and the ACM compressor, exits the secondary heat exchanger. It is directed into the turbine side of the ACM. The steep blade pitch angle of the ACM turbine extracts more energy from the air as it passes through and drives the turbine. Once through, the air is

allowed to expand at the ACM outlet, cooling even further. The combined energy loss from the air first driving the turbine and then expanding at the turbine outlet lowers the system air temperature to near freezing, Fig. 8-2.

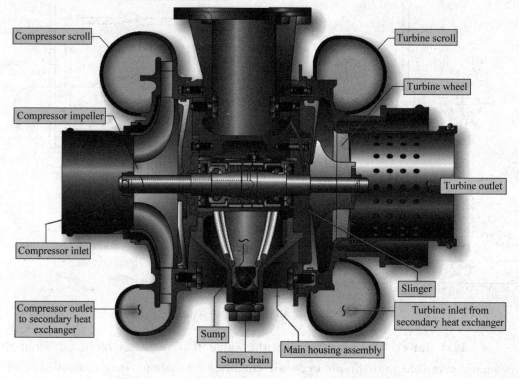

Fig. 8-2　Air cycle machine

[7] **Water Separator**: The cool air from the air cycle machine can no longer hold the quantity of water it could when it was warm. A water separator is used to remove the water from the saturated air before it is sent to the aircraft cabin. The separator operates with no moving parts. Foggy air from the ACM enters and is forced through a fiber glass sock that condenses and coalesces the mist into larger water drops. The convoluted interior structure of the separator swirls the air and water. The water collects on the sides of the separator and drains down and out of the unit, while the dry air passes through. A bypass valve is incorporated in case of a blockage.

[8] **Cabin temperature control system**: Most cabin temperature control systems operate in a similar manner. Temperature is monitored in the cabin, cockpit, conditioned air ducts, and distribution air ducts. These values are input into a temperature controller, or temperature control regulator, normally located in the electronic equipment bay (E/E Bay). A temperature selector in the cockpit can be adjusted to input the desired temperature. The temperature controller compares the actual temperature signals received from the various sensors with the desired temperature input. Circuit logic for the selected mode processes these input signals.

[9] An output signal is sent to a valve in the air cycle air conditioning system. This valve has different names depending on the aircraft manufacturer and design of the environmental control systems (i.e., mixing valve, temperature control valve, trim air valve). It mixes warm bleed air that bypassed the air cycle cooling process with the cold air produced by it. By modulating the valve in response to the signal from the temperature controller, air of the selected temperature is sent to the cabin through the air distribution system.

[10] Cabin temperature pickup units and duct temperature sensors used in the temperature control system are thermistors. Their resistance changes as temperature changes. The temperature selector is a rheostat that varies its resistance as the knob is turned. In the temperature controller, resistances are compared in a bridge circuit. The bridge output feeds a temperature regulating function. An electric signal output is prepared and sent to the valve that mixes hot and cold air. On large aircraft with separate temperature zones, trim air modulating valves for each zone are used. The valves modulate to provide the correct mix required to match the selected temperature. Cabin, flight deck, and duct temperature sensors are strategically located to provide useful information to control cabin temperature.

[11] **Humidification System**: The humidification system which is an optional system increases the level of humidity of the air that flows to the cabin to give the best comfort for the passengers and the crew. The humidification system can be installed to supply the cockpit, some cabin zones. Each humidification zone system is mainly composed of one humidifier module. This humidifier module receives heated air from the mixer of the distribution system, and water from the ATA 38. Water is added to the air to increase its relative humidity and this air flows to its designated zone of the A/C.

[12] **Pressure control system**: Aircraft cabin pressurization can be controlled via two different modes of operation. The first is the isobaric mode, which works to maintain cabin altitude at a single pressure despite the changing altitude of the aircraft. For example, the flight crew may select to maintain a cabin altitude of 8,000 feet (10.92 psi). In the isobaric mode, the cabin pressure is established at the 8,000 feet level and remains at this level, even as the altitude of the aircraft fluctuates.

[13] The second mode of pressurization control is the constant differential mode, which controls cabin pressure to maintain a constant pressure difference between the air pressure inside the cabin and the ambient air pressure, regardless of aircraft altitude changes. The constant differential mode pressure differential is lower than the maximum differential pressure for which the airframe is designed, keeping the integrity of the pressure vessel intact.

[14] When in isobaric mode, the pressurization system maintains the cabin altitude selected by the crew. This is the condition for normal operations. But when the aircraft climbs beyond a certain altitude, maintaining the selected cabin altitude may result in a differential pressure above that for which the airframe was designed. In this case, the

mode of pressurization automatically switches from isobaric to constant differential mode. This occurs before the cabin's max differential pressure limit is reached. A constant differential pressure is then maintained, regardless of the selected cabin altitude.

[15] In addition to the modes of operation described above, the rate of change of the cabin pressure, also known as the cabin rate of climb or descent, is also controlled. This can be done automatically or manually by the flight crew. Typical rates of change for cabin pressure are 300 to 500 fpm. Also, note that modes of pressurization may also refer to "automatic-standby-manual" operation of the pressurization system.

[16] **Pressure regulator**: On many transport category aircraft, two cabin pressure controllers, or a single controller with redundant circuitry, are used. Located in the electronic equipment bay, they receive electric input from the panel selector, as well as ambient and cabin pressure input. Cruise altitude and landing field altitude information are often the crew selection choices on the pressurization control panel. Cabin altitude, rate of climb, and barometric setting are automatic through built-in logic and communication with the air data computer (ADC) and the flight management system (FMS). The controllers process the information and send electric signals to motors that directly position the outflow valve(s).

[17] Modern pressurization control is fully automatic once variable selections are made on the pressurization control panel if, in fact, there are any to be made. Entering or selecting a flight plan into the FMS of some aircraft automatically supplies the pressurization controller with the parameters needed to establish the pressurization schedule for the entire flight. No other input is needed from the crew.

[18] All pressurization systems contain a manual mode that can override automatic control. This can be used in flight or on the ground during maintenance. The operator selects the manual mode on the pressurization control panel. A separate switch is used to position the outflow valve to control cabin pressure.

[19] **Outflow valve**: Controlling cabin pressurization is accomplished through regulating the amount of air that flows out of the cabin. A cabin outflow valve opens, closes, or modulates to establish the amount of air pressure maintained in the cabin. Some outflow valves contain the pressure regulating and the valve mechanism in a single unit. They operate pneumatically in response to the settings on the cockpit pressurization panel that influence the balance between cabin and ambient air pressure.

[20] Pneumatic operation of outflow valves is common. It is simple, reliable, and eliminates the need to convert air pressure operating variables into some other form. Diaphragms, springs, metered orifices, jet pumps, bellows, and poppet valves are used to sense and manipulate cabin and ambient air pressures to correctly position the outflow valve without the use of electricity. Outflow valves that combine the use of electricity with pneumatic operation have all-pneumatic standby and manual modes.

[21] The pressure regulating mechanism can also be found as a separate unit. Many

air transport category aircraft have an outflow valve that operates electrically, using signals sent from a remotely located cabin air pressure controller that acts as the pressure regulator. The controller positions the valve(s) to achieve the settings on the cockpit pressurization panel selectors according to predetermined pressurization schedules. Signals are sent to electric motors to move the valve as needed. On transports, often AC motors are used with a redundant DC motor for standby or manual operations.

[22] **Vapor cycle air conditioning**: is a closed system used solely for the transfer of heat from inside the cabin to outside of the cabin. It can operate on the ground and in flight. This refrigeration cycle is based on the principle that the boiling point of a liquid is raised when the pressure of the vapor around the liquid is raised.

[23] The cycle operates as follows: A liquid refrigerant confined in the receiver at a high pressure is allowed to flow through the expansion valve into the evaporator. The pressure in the evaporator is low enough so that the boiling point of the liquid refrigerant is below the temperature of the air to be cooled, or heat source. Heat flows from the space to be cooled to the liquid refrigerant, causing it to boil (to be converted from liquid to a vapor). Cold vapor from the evaporator enters the compressor, where its pressure is raised, thereby raising the boiling point. The refrigerant at a high pressure and high temperature flows into the condenser.

[24] **Expansion valve**: Refrigerant exits the receiver dryer and flows to the expansion valve. The thermostatic expansion valve has an adjustable orifice through which the correct amount of refrigerant is metered to obtain optimal cooling. This is accomplished by monitoring the temperature of the gaseous refrigerant at the outlet of the next component in the cycle, the evaporator. Ideally, the expansion valve should only let the amount of refrigerant spray into the evaporator that can be completely converted to a vapor, Fig. 8-3.

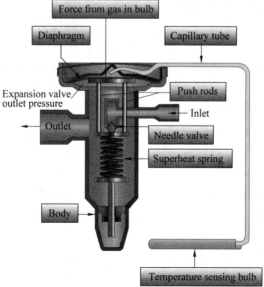

Fig. 8-3　Expansion valve

[25] The temperature of the cabin air to be cooled determines the amount of refrigerant the expansion valve should spray into the evaporator. Only so much is needed to completely change the state of the refrigerant from a liquid to a vapor. Too little causes the gaseous refrigerant to be superheated by the time it exits the evaporator. This is inefficient. Changing the state of the refrigerant from liquid to vapor absorbs much more heat than adding heat to already converted vapor (superheat). The cabin air blowing over the evaporator will not be cooled sufficiently if superheated vapor is flowing through the evaporator. If too much refrigerant is released by the expansion valve into the evaporator, some of it remains liquid when it exits the evaporator. Since it next flows to the compressor, this could be dangerous. The compressor is designed to compress only vapor. If liquid is drawn in and attempts are made to compress it, the compressor could break, since liquids are essentially incompressible.

NEW WORDS AND EXPRESSIONS

(1) anticipate[æn'tɪsɪpeɪt]　　　　　预期，预测
(2) distribute[dɪ'strɪbjuːt]　　　　　分配，分发
(3) stratification[ˌstrætɪfɪ'keɪʃn]　　　成层，分层
(4) layer['leɪə(r)]　　　　　　　　层，膜
(5) humidity[hjuː'mɪdətɪ]　　　　　湿度
(6) fog[fɔg]　　　　　　　　　　雾
(7) panel['pænl]　　　　　　　　面板，壁板
(8) ventilation[ˌventɪ'leɪʃn]　　　　通风
(9) pressurization[ˌpreʃəraɪ'zeɪʃn]　　增压
(10) turbine-powered　　　　　　涡轮驱动，涡轮动力
(11) bleed[bliːd]　　　　　　　　引气
(12) pneumatic[njuː'mætɪk]　　　气动的，气源的
(13) vapor['veɪpə]　　　　　　　蒸发，蒸汽
(14) reciprocating[rɪ'sɪprəkeɪtɪŋ]　　往复式的，活塞式的
(15) automobile['ɔːtəməbiːl]　　　汽车
(16) expansion[ɪk'spænʃn]　　　　膨胀，张开，伸展
(17) turbine['tɜːbaɪn]　　　　　　涡轮
(18) exchanger[ɪks'tʃeɪndʒə]　　　交换器
(19) airflow['eəfləʊ]　　　　　　气流
(20) impeller[ɪm'pelə]　　　　　　叶轮
(21) shaft[ʃɑːft]　　　　　　　　轴
(22) undergo[ˌʌndə'gəʊ]　　　　　遭受，经历
(23) ambient['æmbɪənt]　　　　　周围环境的，周围的

UNIT 8　AIR CONDITIONING SYSTEM

(24) pack[pæk]　　　　　　　　　　　　　组件
(25) manifold['mænɪfəʊld]　　　　　　　总管,歧管
(26) switch[swɪtʃ]　　　　　　　　　　　开关,电门,接通
(27) modulate['mɔdjʊleɪt]　　　　　　　调整,调制
(28) refrigeration[rɪˌfrɪdʒə'reɪʃn]　　　制冷
(29) ram[ræm]　　　　　　　　　　　　冲压
(30) water separator　　　　　　　　　水分离器
(31) saturate['sætʃəreɪt]　　　　　　　浸透的,饱和的
(32) fiberglass sock　　　　　　　　　玻璃纤维凝聚袋
(33) coalesce[ˌkəʊə'les]　　　　　　　结合,聚集
(34) convolute['kɔnvəˌluːt]　　　　　　卷绕,回旋形的
(35) swirl[swəːl]　　　　　　　　　　　涡流,涡旋
(36) bay[beɪ]　　　　　　　　　　　　　舱
(37) thermistor[θəː'mɪstə]　　　　　　热敏电阻
(38) rheostat['riːəˌstæt]　　　　　　　可变电阻器
(39) knob[nɔb]　　　　　　　　　　　　旋钮
(40) bridge[brdʒ]　　　　　　　　　　　电桥
(41) sensor['sensə(r)]　　　　　　　　传感器
(42) strategically[strə'tiːdʒɪklɪ]　　　有策略地
(43) via['vaɪə]　　　　　　　　　　　　经过,通道
(44) isobaric[ˌaɪsəʊ'bærɪk]　　　　　　等压的
(45) fluctuate['flʌktʃʊeɪt]　　　　　　波动,动摇
(46) integrity[ɪn'tegrətɪ]　　　　　　　集成度,完整性
(47) vessel['vesl]　　　　　　　　　　　容器,船舶
(48) intact[ɪn'tækt]　　　　　　　　　　完好无缺的
(49) redundant[rɪ'dʌndənt]　　　　　　多余的,过多的
(50) circuitry['səːkɪtrɪ]　　　　　　　电路
(51) barometric[ˌbærə'metrɪk]　　　　大气压的
(52) ADC: air data computer　　　　大气数据计算机
(53) motor['məʊtə(r)]　　　　　　　　马达
(54) outflow valve　　　　　　　　　　排气活门
(55) parameter[pə'ræmɪtə(r)]　　　　参数
(56) schedule['ʃedjuːl]　　　　　　　　计划,时间表
(57) override[ˌəʊvə'raɪd]　　　　　　　超控
(58) eliminate[ɪ'lɪmɪneɪt]　　　　　　　排除,消除
(59) diaphragm['daɪəfræm]　　　　　　薄膜,隔膜,隔板
(60) orifice['ɔrɪfɪs]　　　　　　　　　　节流孔
(61) jet pump　　　　　　　　　　　　引射泵
(62) bellow['beləʊ]　　　　　　　　　　膜盒,波纹管,风箱

（63）poppet ['pɔpɪt]　　　　　　　　阀芯
（64）manipulate [mə'nɪpjʊleɪt]　　　控制，操纵
（65）manual ['mænjʊəl]　　　　　　手动的，人工的，手册
（66）receiver [rɪ'siːvə(r)]　　　　　制冷剂罐，接收器
（67）evaporator [ɪ'væpə,reɪtə]　　　蒸发器
（68）refrigerant [rɪ'frɪdʒərənt]　　 制冷剂
（69）condenser [kən'densə(r)]　　　冷凝器
（70）thermostatic ['θəːməstætɪc]　　恒温的
（71）optimal ['ɔptɪməl]　　　　　　最佳的

EXERCISE

1. Translate the followings into Chinese.

（1）**Cabin altitude**. Given the air pressure inside the cabin, the altitude on a standard day that has the same pressure as that in the cabin. Rather than saying the pressure inside the cabin is 10.92 psi, it can be said that the cabin altitude is 8,000 feet (mean sea level, MSL).

（2）**Cabin differential pressure**. The difference between the air pressure inside the cabin and the air pressure outside the cabin (psid or Δpsi).

（3）**Cabin rate of climb**. The rate of change of air pressure inside the cabin, expressed in feet per minute (fpm) of cabin altitude change.

（4）**Aircraft altitude**. The actual height above sea level at which the aircraft is flying.

（5）**Ambient pressure**. The pressure in the area immediately surrounding the aircraft.

（6）**Manifold**. A duct or chamber with several inlets and outlets for air fluid distribution.

（7）**Receiver-dryer**. The component in the high pressure side of a vapor-cycle cooling system that serves as a reservoir for the liquid refrigerant. The receiver-dryer contains a desiccant that absorbs any moisture that may be in the refrigerant.

（8）**Evaporator**. The component in a vapor cycle air conditioning system in which the refrigerant absorbs heat from the aircraft cabin. The refrigerant in the evaporator coils is a liquid in the form of tiny droplets. Warm air from the cabin blows through the fins on the coils, and heat from this air is absorbed by the refrigerant, changing the droplets of liquid into a vapor. The heat used to make this change is taken from the air and as a result the air passing through the evaporator is cooled.

（9）**ADC**. An electronic computer in an aircraft that senses pitot pressure, static pressure, and total air temperature. It produces an indication of altitude, indicated airspeed, true airspeed, and Mach number. The output of the ADC is usable by any of the engine or flight control computers.

(10) **Switch**. A component used to open or close an electrical circuit or to select paths through the circuit.

2. Answer the following questions in English.
(1) What is the working principle of air cycle cooling system?
(2) What is the working principle of vapor cycle cooling system?
(3) What is the working principle of cabin temperature control system?
(4) What is the function of pack valve?
(5) Which parameters are controlled by the pressure control system?
(6) What is the function of expansion valve?
(7) Which part(s) of aircraft are pressurized by the air conditioning system?
(8) Are there any un-pressurized areas in aircraft? Where are they?

SUPPLEMENTARY READING

[1] **Primary Heat Exchanger:** Generally, the warm air dedicated to pass through the air cycle system first passes through a primary heat exchanger. It acts similarly to the radiator in an automobile. A controlled flow of ram air is ducted over and through the exchanger, which reduces the temperature of the air inside the system.

[2] A fan draws air through the ram air duct when the aircraft is on the ground so that the heat exchange is possible when the aircraft is stationary. In flight, ram air doors are modulated to increase or decrease ram air flow to the exchanger according to the position of the wing flaps. During slow flight, when the flaps are extended, the doors are open. At higher speeds, with the flaps retracted, the doors move toward the closed position reducing the amount of ram air to the exchanger. Similar operation is accomplished with a valve on smaller aircraft.

[3] **Evaporator:** Most evaporators are constructed of copper or aluminum tubing coiled into a compact unit. Fins are attached to increase surface area, facilitating rapid heat transfer between the cabin air blown over the outside of the evaporator with a fan and the refrigerant inside. The expansion valve located at the evaporator inlet releases high-pressure, high-temperature liquid refrigerant into the evaporator. As the refrigerant absorbs heat from the cabin air, it changes into a low-pressure vapor. This is discharged from the evaporator outlet to the next component in the vapor cycle system, the compressor. The temperature and pressure pickups that regulate the expansion valve are located at the evaporator outlet.

[4] The evaporator is situated in such a way that cabin air is pulled to it by a fan.

The fan blows the air over the evaporator and discharges the cooled air back into the cabin. This discharge can be direct when the evaporator is located in a cabin wall. A remotely located evaporator may require ducting from the cabin to the evaporator and from the evaporator back into the cabin. Sometimes the cool air produced may be introduced into an air distribution system where it can blow directly on the occupants through individual delivery vents. In this manner, the entire vapor cycle air conditioning system may be located fore or aft of the cabin. A multi position fan switch controlled by the pilot is usually available.

[5] When cabin air is cooled by flowing over the evaporator, it can no longer retain the water that it could at higher temperature. As a result, it condenses on the outside of the evaporator and needs to be collected and drained overboard. Pressurized aircraft may contain a valve in the evaporator drain line that opens only periodically to discharge the water, to maintain pressurization. Fins on the evaporator must be kept from being damaged, which could inhibit airflow. The continuous movement of warm cabin air around the fins keeps condensed water from freezing. Ice on the evaporator reduces the efficiency of the heat exchange to the refrigerant.

[6] **Compressor**: The compressor is the heart of the vapor cycle air conditioning system. It circulates the refrigerant around the vapor cycle system. It receives low-pressure, low-temperature refrigerant vapor from the outlet of the evaporator and compresses it. As the pressure is increased, the temperature also increases. The refrigerant temperature is raised above that of the outside air temperature. The refrigerant then flows out of the compressor to the condenser where it gives off the heat to the outside air.

[7] The compressor is the dividing point between the low side and the high side of the vapor cycle system. Often it is incorporated with fittings or has fittings in the connecting lines to it that are designed to service the system with refrigerant. Access to the low and high sides of the system are required for servicing, which can be accomplished with fitting upstream and downstream of the compressor.

[8] Modern compressors are either engine driven or driven by an electric motor. Occasionally, a hydraulically driven compressor is used. A typical engine-driven compressor, similar to that found in an automobile, is located in the engine nacelle and operated by a drive belt off the engine crankshaft. An electromagnetic clutch engages when cooling is required, which causes the compressor to operate. When cooling is sufficient, power to the clutch is cut, and the drive pulley rotates but the compressor does not.

[9] Dedicated electric motor-driven compressors are also used on aircraft. Use of an electric motor allows the compressor to be located nearly anywhere on the aircraft, since wires can be run from the appropriate bus to the control panel and to the compressor. Hydraulically-driven compressors are also able to be remotely located. Hydraulic lines from the hydraulic manifold are run through a switch activated solenoid to the compressor. The solenoid allows fluid to the compressor or bypasses it. This controls the operation of the hydraulically

driven compressor.

[10] **Condenser**: The condenser is the final component in the vapor cycle. It is a radiator-like heat exchanger situated so that outside air flows over it and absorbs heat from the high-pressure, high temperature refrigerant received from the compressor. A fan is usually included to draw the air through the compressor during ground operation. On some aircraft, outside air is ducted to the compressor. On others, the condenser is lowered into the airstream from the fuselage via a hinged panel. Often, the panel is controlled by a switch on the throttle levers. It is set to retract the compressor and streamline the fuselage when full power is required.

[11] The outside air absorbs heat from the refrigerant flowing through the condenser. The heat loss causes the refrigerant to change state back into a liquid. The high-pressure liquid refrigerant then leaves the condenser and flows to the receiver dryer. A properly engineered system that is functioning normally fully condenses all the refrigerant flowing through the condenser.

[12] In normal pack operation the purpose of the anti-ice valve is to prevent ice blockage across the condenser. If the pack controller is unable to control the bypass valve the anti-ice valve is signaled to pneumatically control the pack outlet temperature to 15℃ (59.0°F). The valve stops the build-up of ice across the condenser using a differential pressure regulator. This differential pressure regulator opens the valve and delivers hot air to the condenser. Port senses condenser inlet high-pressure and port senses condenser outlet high-pressure. When ice builds up along the flow path through the condenser the pressure drop increases rapidly. The differential pressure between the ports increases. This opens the poppet valve which allows a high flow of muscle pressure into the pneumatic actuator, which opens the anti-ice valve.

TOPIC 8：单位及单位换算

在民航业用到的单位术语经常采用缩写,各单位术语缩写和释义如表 T8-1 所示。

表 T8-1　单位缩写及含义

缩　写	术　语	释　义
dB	decibel	分贝
dB(A)	a-weighted decibel	A 分贝（噪声）
ft	feet, foot	英尺
ft/min	feet per minute	每分钟英尺
gal	gallon	加仑
in	inch	英寸
kg	kilogram	千克
kg/m^3	kilograms/cubic meter	千克/立方米
kPa	kilo pascal	千帕
kW	kilowatt	千瓦
lb	pound	磅
psi	pounds per square inch	每平方英寸磅力
mbar	millibar	毫巴
pH	acidity or alkalinity	酸碱性
℃	degrees centigrade	摄氏度
℉	degrees fahrenheit	华氏度
K	degrees kelvin	开氏度
R	degrees rankine	兰氏度
rpm	revolution per minute	每分钟转数

在民航业,航空器制造商和使用维修企业通常采用英制单位进行度量,它们与公制单位的换算如表 T8-2～表 T8-4 所示。

表 T8-2　公制度量和非公制度量之间的换算关系

量的名称	British System(英制)		Metric(公制)
length （长度）	1inch(in)（英寸）		= 25.4 millimetres（毫米）
	12 inches（英寸）	= 1 foot(ft)（英尺）	= 30.48 centimetres（厘米）
	3 feet（英尺）	= 1 yard(yd)（码）	= 0.914 metre（米）
	2,025.8 yards（码）	= 1 n mile（海里）	= 1.852 kilometres（公里）
	1,760 yards（码）	= 1 mile（英里）	= 1.609 kilometres（公里）
area （面积）	1 square(sq)inch （平方英寸）		= 6.452 sq centimetres(cm^2) （平方厘米）
	144 sq inches（平方英寸）	= 1 sq foot（平方英尺）	= 929.03 sq centimetres （平方厘米）
	9 sq feet（平方英尺）	= 1 sq yard（平方码）	= 0.836 sq metre（平方米）
	4,840 sq yards（平方码）	= 1 acre（英亩）	= 0.405 hectare（公顷）

续表

量的名称	British System(英制)		Metric(公制)
weight （重量）	437 grains(格令)	=1 ounce(oz)(盎司)	=28.35 grams(克)
	16 ounces(盎司)	=1 pound(lb)(磅)	=0.454 kilogram(公斤)
	14 pounds(磅)	=1 stone(st)(英石)	=6.356 kilograms(公斤)
British capacity （英国容量）	20 fluid ounces (fl oz)(液盎司)	=1 pint(pt)(品脱)	=0.568 litre(升)
	2 pints(品脱)	=1 quart(qt)(夸脱)	=1.136 litres(升)
	8 pints(品脱)	=1 gallon(gal)(加仑)	=4.546 litres(升)
American capacity （美国容量）	16 US fluid ounces （美国液盎司）	=1 US pint(美国品脱)	=0.473 litre
	2 US pints(美国品脱)	=1 US quart(美国夸脱)	=0.946 litre(升)
	4 US quarts(美国夸脱)	=1 US gallon(美国加仑)	=3.785 litre(升)

表 T8-3 压力单位的换算表

标准大气压(atm)	psi	Pa	bar	mm 汞柱	m 水柱
1	14.7	101,325	1.013,25	760	10.336

表 T8-4 速度单位的换算表

米/秒 m/s	公里/小时 (km/h)	海里/小时 (nm/h)	节 (knot)	英里/小时 (MPH)	码/秒 (yard/s)
1	3.6	1.94	1.94	2.236,9	1.093,6

UNIT 9

FLIGHT CONTROL SYSTEM

[1] Flight control is the action taken to make the aircraft follow any desired flight path. When an aircraft is said to be controllable, it means that the craft responds easily and promptly to movement of the controls. Different control surfaces are used to control the aircraft about each of the three axes. Moving the control surfaces on an aircraft changes the airflow over the aircraft's surface. This, in turn, creates changes in the balance of forces acting to keep the aircraft flying straight and level. The flight control surfaces are hinged or movable airfoils designed to change the attitude of the aircraft during flight. These surfaces are illustrated in Fig. 9-1. Aircraft flight control systems consist of primary and secondary systems.

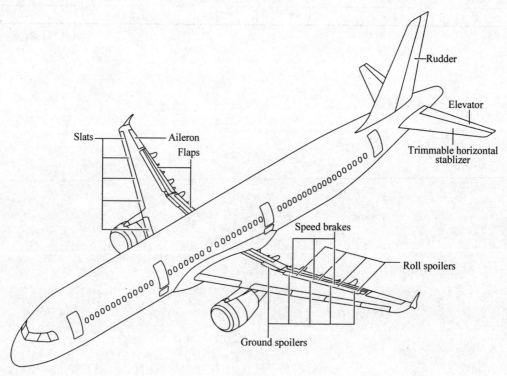

Fig. 9-1　A typical flight control system

[2] The primary flight control surfaces on a fixed-wing aircraft include: ailerons, elevators, and rudder, are required to control an aircraft safely during flight. The ailerons are attached to the trailing edge of both wings and when moved, rotate the aircraft around the longitudinal axis. The elevators are attached to the trailing edge of the horizontal stabilizer. When moved, they alter aircraft pitch, which is the attitude about the horizontal or lateral axis. The rudder is hinged to the trailing edge of the vertical stabilizer. When the rudder position changes, the aircraft rotates about the vertical axis (yaw). Fig.9-2 shows the primary flight controls of a light aircraft and the movement they create relative to the three axes of flight.

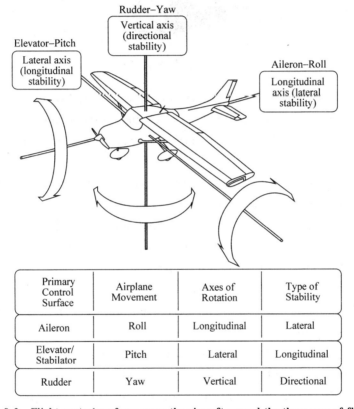

Primary Control Surface	Airplane Movement	Axes of Rotation	Type of Stability
Aileron	Roll	Longitudinal	Lateral
Elevator/ Stabilator	Pitch	Lateral	Longitudinal
Rudder	Yaw	Vertical	Directional

Fig. 9-2　Flight control surfaces move the aircraft around the three axes of flight

[3] **Ailerons:** Ailerons are the primary flight control surfaces that move the aircraft about the longitudinal axis. In other words, movement of the ailerons in flight causes the aircraft to roll. Ailerons are usually located on the outboard trailing edge of each wing. They are built into the wing and are calculated as part of the wing's surface area.

[4] Ailerons are controlled by a side-to-side motion of the control stick in the cockpit or a rotation of the control yoke. When the aileron on one wing deflects down, the aileron on the opposite wing deflects upward. This amplifies the movement of the aircraft around the longitudinal axis. On the wing on which the aileron trailing edge moves downward, camber is increased, and lift is increased. Conversely, on the other

wing, the raised aileron decreases lift, as shown in Fig. 9-3. The result is a sensitive response to the control input to roll the aircraft.

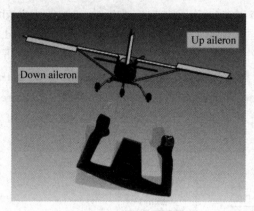

Fig. 9-3   Aileron control

[5] The pilot's request for aileron movement and roll are transmitted from the cockpit to the actual control surface in a variety of ways depending on the aircraft. A system of control cables and pulleys, push-pull tubes, hydraulics, electric, or a combination of these can be employed.

[6] **Elevator**: The elevator is the primary flight control surface that moves the aircraft around the horizontal or lateral axis. This causes the nose of the aircraft to pitch up or down. The elevator is hinged to the trailing edge of the horizontal stabilizer and typically spans most or all of its width. It is controlled in the cockpit by pushing or pulling the control stick or yoke forward or aft.

[7] Light aircraft use a system of control cables and pulleys or push-pull tubes to transfer cockpit inputs to the movement of the elevator. High-performance and large aircraft typically employ more complex systems. Hydraulic power is commonly used to move the elevator on these aircraft. On aircraft equipped with fly-by-wire controls, a combination of electrical and hydraulic power is used.

[8] **Rudder**: The rudder is the primary control surface that causes an aircraft to yaw or move about the vertical axis. This provides directional control and thus points the nose of the aircraft in the direction desired. Most aircraft have a single rudder hinged to the trailing edge of the vertical stabilizer. It is controlled by a pair of foot-operated rudder pedals in the cockpit. When the right pedal is pushed forward, it deflects the rudder to the right which moves the nose of the aircraft to the right. The left pedal is rigged to simultaneously move aft. When the left pedal is pushed forward, the nose of the aircraft moves to the left.

[9] As with the other primary flight controls, the transfer of the movement of the cockpit controls to the rudder varies with the complexity of the aircraft. Many aircraft incorporate the directional movement of the nose or tail wheel into the rudder control

system for ground operation. This allows the operator to steer the aircraft with the rudder pedals during taxi when the airspeed is not high enough for the control surfaces to be effective. Some large aircraft have a split rudder arrangement. This is actually two rudders, one above the other. At low speeds, both rudders deflect in the same direction when the pedals are pushed. At higher speeds, one of the rudders becomes inoperative as the deflection of a single rudder is aerodynamically sufficient to maneuver the aircraft.

[10] **Secondary or auxiliary control surfaces**: Wing flaps, leading edge devices, spoilers, and trim systems constitute the secondary control system and improve the performance characteristics of the airplane or relieve the pilot of excessive control forces. There are several secondary or auxiliary flight control surfaces. Their names, locations, and functions of those for most large aircraft are listed in Table 9-1.

Table 9-1 Secondary or auxiliary control surfaces and respective locations for a large aircraft

Secondary/Auxiliary Flight Control Surfaces		
Name (Item)	Location	Function
Flap	Inboard trailing edge of wings	Extends the camber of the wing for greater lift and slower flight. Allows control at low speeds for short field takeoffs and landings.
Trim tab	Trailing edge of primary flight control surfaces	Reduces the force needed to move a primary control surface.
Balance tab	Trailing edge of primary flight control surfaces	Reduces the force needed to move a primary control surface.
Anti-balance tab	Trailing edge of primary flight control surfaces	Increases feel and effectiveness of primary control surface.
Servo tab	Trailing edge of primary flight control surfaces	Assists or provides the force for moving a primary flight control.
Spoiler	Upper and trailing edge of wing	Decreases (spoils) lift. Can augment aileron function.
Slat	Mid to outboard leading edge of wing	Extends the camber of the wing for greater lift and slower flight. Allows control at low speeds for short field takeoffs and landings.
Slot	Outer leading edge of wing forward of ailerons	Directs air over upper surface of wing during high angle of attack. Lowers stall speed and provides control during slow flight.
Leading edge flap	Inboard leading edge of wing	Extends the camber of the wing for greater lift and slower flight. Allows control at low speeds for short field takeoffs and landings.
NOTE: An aircraft may possess none, one, or a combination of the above control surfaces.		

[11] **Flaps**: Flaps are found on most aircraft. They are usually inboard on the wings' trailing edges adjacent to the fuselage. Leading edge flaps are also common. They extend forward and down from the inboard wing leading edge. The flaps are lowered to increase the camber of the wings and provide greater lift and control at slow speeds. They enable landing at slower speeds and shorten the amount of runway required for

takeoff and landing. The amount that the flaps extend and the angle they form with the wing can be selected from the cockpit. Typically, flaps can extend up to 45°-50°.

[12] **Leading edge devices**: High-lift devices also can be applied to the leading edge of the airfoil. The most common types are fixed slots, movable slats, leading edge flaps, and cuffs (Fig. 9-4).

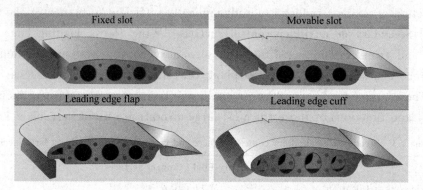

Fig. 9-4 Leading edge high-lift devices

[13] **Spoilers**: Found on many gliders and some aircraft, high drag devices called spoilers are deployed from the wings to spoil the smooth airflow, reducing lift and increasing drag. On gliders, spoilers are most often used to control rate of descent for accurate landings. On other aircraft, spoilers are often used for roll control, an advantage of which is the elimination of adverse yaw. To turn right, for example, the spoiler on the right wing is raised, destroying some of the lift and creating more drag on the right. The right wing drops, and the aircraft banks and yaws to the right. Deploying spoilers on both wings at the same time allows the aircraft to descend without gaining speed. Spoilers are also deployed to help reduce ground roll after landing. By destroying lift, they transfer weight to the wheels, improving braking effectiveness.

[14] Fig. 9-5 shows the central control mechanism of a transport airplane. The control columns and control wheels are controlled by pilots' hands, and the pedals are controlled by pilots' feet. When the pilots push or pull the control wheels and columns, the elevators will be moved downward or upward respectively. When the pilots turn the control wheels to left, the left aileron will be moved upward and the right aileron will be moved downward. When the control wheels are turned to right, the left aileron will be moved downward, and the right aileron will be moved upward. In some smaller airplanes, a control stick is equipped instead of the control wheel and column. The rudder will be moved to right if the right foot pedal is moved forward and the left foot pedal is moved rearward (Fig. 9-6).

[15] The transmission systems connect the control surfaces to the central control mechanism in the cockpit. Two kinds of mechanical linkage are used in flight control systems. They are the cable-type linkage and the push-pull rod-type linkage. For high

speed and large aircraft, hydraulic or electrical power assist control system is equipped in the transmission system to overcome the great airflow resistance to control movement.

Fig. 9-5　Control wheel mechanism

Fig. 9-6　Rudder control mechanism

[16] **Flight control system**: A control system is a collection of mechanical and electronic equipment that allows an aircraft to be flown with exceptional precision and reliability. A control system consists of cockpit controls, sensors, actuators (hydraulic, mechanical or electrical) and computers. There are two types used in modern aircraft: mechanical and fly-by-wire.

[17] **Mechanical flight control system**: Mechanical or manually operated flight control systems are the most basic method of controlling an aircraft, and used in early aircraft and currently in small aircraft where the aerodynamic forces are not excessive. This system gives pilots a lot of "feel" as he is directly connected to the control surfaces on the aircraft. The system is generally made up of cables, pulleys, rods, and sometimes even chains to transmit the forces applied from the cockpit controls directly to the control surfaces, as shown in Fig. 9-7.

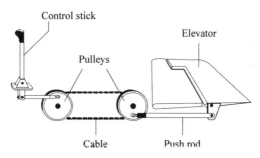

Fig. 9-7　Mechanical control system

[18] Mechanical flight control system is not very good when there are large stick forces. The pilot's strength becomes the limiting factors. Thus, some mechanical flight control systems use servo tabs that provide aerodynamic assistance. Servo tabs are small surfaces hinged to the control surfaces. The flight control mechanisms move these tabs, aerodynamic forces in turn move, or assist the movement of the control surfaces

reducing the amount of mechanical forces needed. This arrangement was used in early piston-engined transport aircraft and in early jet transports. In large aircraft, the control surfaces are operated by hydraulic actuators controlled by valves moved by control yoke and rudder pedals.

[19] **Fly-by-wire (FBW)** is the generally accepted term for flight control systems in which a computer processes the pilot's control movements and sends electric signals to the flight control surface actuators without any mechanical linkage. It is a computer-configured control, where a computer system is interposed between the operator and the final control actuators or surfaces. It modifies the manual inputs of the pilot in accordance with control parameters.

[20] The FBW architecture was developed in 1970's. The supersonic Concorde can be considered a first and isolated civil aircraft equipped with fly-by-wire system. In the 80's the digital technique was imported from military into civil aviation by Airbus, first with the A320, then followed by A330/A340, B777 and A380.

[21] In a fly-by-wire system, computer technology is used in the operation of the primary flight controls. There is no direct linkage between the pilot and the control surfaces (Fig. 9-8). Control column movement inputs control demands into a series of computers. Computer software compares pilot's control inputs to aircraft's control parameters (airspeed and angle of attack). Software send command signal(s) to the appropriate flight controls. Command signals are routed via electrical cables to a servo valve attached to the hydraulic power unit. Signal is converted into servo valve displacement, allowing the hydraulic flow and pressure to actuate the control surface.

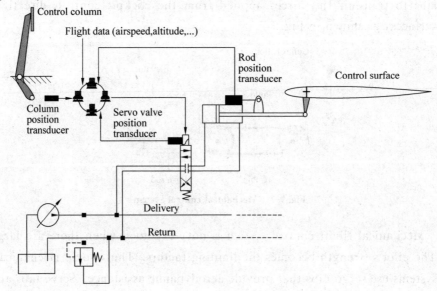

Fig. 9-8 FBW control system

NEW WORDS AND EXPRESSIONS

(1) axis['æksɪːs]　　　　　　　　　　　　轴,坐标轴
(2) hinge[hɪndʒ]　　　　　　　　　　　　铰接,铰链
(3) aileron['eɪlərɔn]　　　　　　　　　　　副翼
(4) elevator['elɪveɪtə]　　　　　　　　　　升降舵
(5) rudder['rʌdə(r)]　　　　　　　　　　 方向舵
(6) trailing edge　　　　　　　　　　　　 后缘
(7) pitch[pɪtʃ]　　　　　　　　　　　　　 俯仰
(8) yaw[jɔː]　　　　　　　　　　　　　　 偏航
(9) roll[rəʊl]　　　　　　　　　　　　　　滚转
(10) control stick　　　　　　　　　　　　操纵杆,驾驶杆
(11) control yoke　　　　　　　　　　　　驾驶盘
(12) cable['keɪbl]　　　　　　　　　　　　钢索,电缆
(13) pulley['pʊlɪ]　　　　　　　　　　　　滑车,滑轮
(14) push-pull tube　　　　　　　　　　　推拉杆
(15) pedal['pedl]　　　　　　　　　　　　脚蹬,踏板
(16) split[splɪt]　　　　　　　　　　　　　分离,开裂
(17) flap[flæp]　　　　　　　　　　　　　襟翼
(18) spoiler['spɔɪlə(r)]　　　　　　　　　扰流板
(19) trim tab　　　　　　　　　　　　　　配平调整片
(20) balance tab　　　　　　　　　　　　平衡调整片
(21) anti-balance tab　　　　　　　　　　反平衡调整片
(22) servo tab　　　　　　　　　　　　　伺服调整片
(23) slat[slæt]　　　　　　　　　　　　　缝翼
(24) slot[slɔt]　　　　　　　　　　　　　开缝,翼缝
(25) adverse['ædvəːs]　　　　　　　　　　逆的,相反的
(26) bank[bæŋk]　　　　　　　　　　　　使倾斜,带坡度转弯
(27) control column　　　　　　　　　　　驾驶杆
(28) control wheel　　　　　　　　　　　 驾驶盘
(29) transmission[træns'mɪʃn]　　　　　　传动
(30) linkage['lɪŋkɪdʒ]　　　　　　　　　　连杆,机械连接
(31) fly-by-wire　　　　　　　　　　　　 电传操纵
(32) interposed[ˌɪntə'pəʊzd]　　　　　　　介入,插入

EXERCISE

1. Translate the followings into Chinese.

(1) **Cuff**. Specially shaped pieces of sheet metal attached to the leading edge of the wing to increase the camber, or curvature, of the airfoil to improve the slow-flight characteristics of the wing.

(2) **Aileron**. A primary flight control surface mounted on the trailing edge of an airplane wing, near the tip. Ailerons operate by lateral movement of the control wheel or stick, and their displacement causes the airplane to rotate about its longitudinal axis.

(3) **Elevator**. The horizontal, movable control surface in the tail section, or empennage, of an airplane. The elevator is hinged to the trailing edge of the fixed horizontal stabilizer. Moving the elevator up or down, by fore-and-aft movement of the control yoke or stick, changes the aerodynamic force produced by the horizontal tail surface.

(4) **Rudder**. The movable control surface mounted on the trailing edge of the vertical fin of an airplane. The rudder is moved by foot-operated pedals in the cockpit, and movement of the rudder rotates the airplane about its vertical axis.

(5) **Rudder vator**. Control surfaces on an airplane that combine the functions of the rudder and the elevators. Rudder vators are the movable surfaces in a V-tail empennage. Moving together, they act as elevators, rotating the airplane about its lateral axis. Moving differentially, they act as the rudder to rotate the airplane about its vertical axis.

(6) **Leading edge**. The edge of a moving object that reaches a point in space or time ahead of the rest of the object. In an airplane wing or a helicopter rotor, the leading edge is the part of the wing or rotor the moving air touches first. In a pulse of electrical energy, the leading edge is the first part of the pulse that moves away from the quiescent, or at-rest, state.

(7) **Trailing edge**. The back edge of an airfoil, such as a wing, a helicopter rotor, or a propeller blade. It is the edge that passes through the air last.

(8) **Spoiler**. A small deflector that can be raised into the air flowing over an airfoil. When a spoiler is raised into the airstream, it disturbs the smooth flow of air and destroys part of the lift the airfoil is producing. Spoilers are used on high-performance sailplanes to decrease the lift so they will not float during landing.

(9) **Tab**. A small, movable control hinged to the trailing edge of one of an airplane's primary flight control surfaces. Tabs can be used to help the pilot move the primary control surface, or they can be used to produce an aerodynamic load on the primary surface that trims the aircraft for a hands-off flight condition. eg. antiservo tab, balance tab, servo tab, spring tab, trim tab.

(10) **Slot.** A fixed, nozzle-like opening near the leading edge of an airplane wing, ahead of the aileron. A slot acts as a duct to force air down on the upper surface of the wing when the airplane is flying at a high angle of attack. Forcing the air down on the top of the wing allows the airplane to fly at a higher angle of attack before it stalls. Since the slot is located ahead of the aileron, the aileron remains effective even at very high angles of attack.

2. **Answer the following questions in English.**
(1) What does it mean by saying that "an aircraft is controllable"?
(2) What happens when the control surfaces on an aircraft are moved?
(3) What is the function of ailerons?
(4) What is the function of rudder?
(5) What is the function of elevators?
(6) Which surfaces are included in auxiliary control surfaces groups?
(7) What is the function of the transmission systems?
(8) What are the technical characteristics of the fly-by-wire system?

SUPPLEMENTARY READING

[1] The ailerons, elevators, and rudder are considered conventional primary control surfaces. However, some aircraft are designed with a control surface that may serve a dual purpose. For example, elevons perform the combined functions of the ailerons and the elevator.

[2] A movable horizontal tail section, called a stabilator, is a control surface that combines the action of both the horizontal stabilizer and the elevator. Basically, a stabilator is a horizontal stabilizer that can also be rotated about the horizontal axis to affect the pitch of the aircraft.

[3] A ruddervator combines the action of the rudder and elevator. This is possible on aircraft with V-tail empennages where the traditional horizontal and vertical stabilizers do not exist. Instead, two stabilizers angle upward and outward from the aft fuselage in a "V" configuration. Each contains a movable ruddervator built into the trailing edge. Movement of the ruddervators can alter the movement of the aircraft around the horizontal and/or vertical axis. Additionally, some aircraft are equipped with flaperons. Flaperons are ailerons which can also act as flaps.

[4] There are various kinds of flaps. Plain flaps form the trailing edge of the wing when the flap is in the retracted position. The airflow over the wing continues over the upper and lower surfaces of the flap, making the trailing edge of the flap essentially the trailing edge of the wing. The plain flap is hinged so that the trailing edge can be lowered. This increases wing camber and provides greater lift.

[5] A split flap is normally housed under the trailing edge of the wing. It is usually

just a braced flat metal plate hinged at several places along its leading edge. The upper surface of the wing extends to the trailing edge of the flap. When deployed, the split flap trailing edge lowers away from the trailing edge of the wing. Airflow over the top of the wing remains the same. Airflow under the wing now follows the camber created by the lowered split flap, increasing lift.

[6] Fowler flaps not only lower the trailing edge of the wing when deployed but also slide aft, effectively increasing the area of the wing. This creates more lift via the increased surface area, as well as the wing camber. When stowed, the fowler flap typically retracts up under the wing trailing edge similar to a split flap. The sliding motion of a fowler flap can be accomplished with a worm drive and flap tracks.

[7] An enhanced version of the fowler flap is a set of flaps that actually contains more than one aerodynamic surface. Fig. 9-9 shows a triple-slotted flap. In this configuration, the flap consists of a fore flap, a mid flap, and an aft flap. When deployed, each flap section slides aft on tracks as it lowers. The flap sections also separate leaving an open slot between the wing and the fore flap, as well as between each of the flap sections. Air from the underside of the wing flows through these slots. The result is that the laminar flow on the upper surfaces is enhanced. The greater camber and effective wing area increase overall lift.

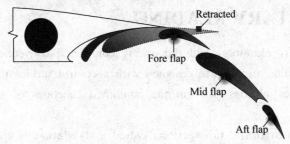

Fig. 9-9　Triple-slotted flap

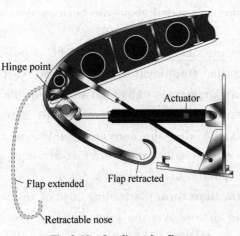

Fig. 9-10　Leading edge flap

[8] Heavy aircraft often have leading edge flaps that are used in conjunction with the trailing edge flaps, Fig. 9-10. They can be made of machined magnesium or can have an aluminum or composite structure. While they are not installed or operate independently, their use with trailing edge flaps can greatly increase wing camber and lift. When stowed, leading edge flaps retract into the leading edge of the wing.

[9] The differing designs of leading edge flaps essentially provide the same effect. Activation of the trailing edge flaps automatically deploys the leading edge flaps, which are driven out of the leading edge and downward, extending the camber of the wing.

TOPIC 9：字母与数字特殊读法

在航空业中，为了避免沟通中出现差错，英文字母和数字在一些特殊场合有其特殊的读法。

1）航空无线电 26 英文字母读法

在无线电通信中，为了避免由于传输质量或口音问题发生误听差错，每一个英文字母一般用一个约定俗成的单词表示。根据国际电信联盟（International Telecommunication Union，ITU）的规定，民用通信中应运用国际民航组织（International Civil Aviation Organization，ICAO）使用的解释法作为其"标准解释法"，具体见表 T9-1。军方也经常用这套规则，可以看看美国电影里都是这样的：如通过对讲发出战斗机攻击指令密码的时候。

表 T9-1 英文字母在航空无线电领域的读法

字母	读法	音标	字母	读法	音标
A	Alfa	[ælfɑ]	N	November	[nɔvembə]
B	Bravo	[brɑːvɔ]	O	Osacr	[ɔskɑ]
C	Charlie	[tʃɑːli]	P	Papa	[pəpɑ]
D	Delta	[deltɑ]	Q	Quebec	[kwəˈbɛk]
E	Echo	[ekɔ]	R	Romeo	[rɔːmiɔ]
F	Foxtrot	[fɔkstrɔt]	S	Sierra	[sierɑ]
G	Golf	[gʌlf]	T	Tango	[tæŋɔ]
H	Hotel	[hɔːtel]	U	Uniform	[juːnifɔːm]
I	India	[indiɑ]	V	Victor	[viktɑ]
J	Juliet	[dʒuːlie]	W	Whisky	[wiski]
K	Kilo	[kiːlɔ]	X	X-ray	[eks rei]
L	Lima	[liːmɑ]	Y	Yankee	[jænki]
M	Mike	[maik]	Z	Zulu	[zuːluː]

2）数字发音

数字发音具体见表 T9-2。

表 T9-2 数字发音

阿拉伯数字	0	1	2	3	4	5	6	7	8	9
中文读法	零	一	二	三	四	五	六	七	八	九
我国无线电读法	洞	幺	两	三	四	五	六	拐	八	勾
英文读法	Zero	One	Two	Three	Four	Five	Six	Seven	Eight	Nine

3）希腊字母读法

希腊字母读法具体见表 T9-3。

表 T9-3　希腊字母读法

字母	音标注音	中文注音	字母	音标注音	中文注音
α	alfa	阿尔法	μ	miu	缪
β	beta	贝塔	ν	niu	纽
γ	gamma	伽马	ξ	ksi	柯西
δ	delta	德尔塔	ο	omikron	奥密克戎
ε	epsilon	埃普西隆	π	pai	派
ζ	zeta	截塔	ρ	rou	柔
η	eta	艾塔	σ	sigma	西格玛
θ	θita	西塔	τ	tau	套
ι	iota	约塔	υ	jupsilon	宇普西龙
κ	kappa	卡帕	φ	fai	佛爱
λ	lambda	兰姆达	χ	khai	凯
ψ	psai	普赛	ω	omiga	欧米伽

UNIT 10

AIRCRAFT FUEL SYSTEM

[1] The aircraft fuel system is designed to provide an uninterrupted flow of clean fuel from the fuel tanks to the engine. The fuel must be available to the engine under all conditions of engine power, altitude, attitude, and during all approved flight maneuvers. A fuel system consists of storage tanks, pumps, filters, valves, fuel lines, metering devices, and monitoring devices. Each fuel system must be designed and arranged to provide independence between multiple fuel storage and supply systems so that failure of any one component in one system will not result in loss of fuel storage or supply of another system.

[2] The fuel system must be designed and arranged to prevent the ignition of the fuel within the system by direct lightning strikes or swept lightning strokes to areas where such occurrences are highly probable, or by corona or streamering at fuel vent outlets. A corona is a luminous discharge that occurs as a result of an electrical potential difference between the aircraft and the surrounding area. Streamering is a branch-like ionized path that occurs in the presence of a direct stroke or under conditions when lightning strikes are imminent.

[3] Aircraft with turbine engines use a type of fuel different from that of reciprocating aircraft engines. Commonly known as jet fuel, turbine engine fuel is designed for use in turbine engines and should never be mixed with aviation gasoline (AVGAS) or introduced into the fuel system of a reciprocating aircraft engine.

[4] **Fuel**: Three basic turbine engine fuel types are available worldwide, although some countries have their own unique fuels. The first is Jet A. It is the most common turbine engine fuel available in the continental United States. Globally, Jet A-1 is the most popular. Both Jet A and Jet A-1 are fractionally distilled in the kerosene range. They have low volatility and low vapor pressure. Flashpoints range between 110°F and 150°F. Jet A freezes at $-40°F$ and Jet A-1 freezes at $-52.6°F$. Most engine operation manuals permit the use of either Jet A or Jet A-1. The third basic type of turbine engine fuel available is Jet B. It is a wide-cut fuel that is basically a blend of kerosene and gasoline. Its volatility and vapor pressure reflect this and fall between Jet A and AVGAS. Jet B is primarily available in Alaska and Canada due to its low freezing point

of approximately -58°F, and its higher volatility yields better cold weather performance.

[5] **Fuel tank**: There are three basic types of aircraft fuel tank: rigid removable tank, bladder tank, and integral fuel tank. The type of aircraft, its design and intended use, as well as the age of the aircraft determine which fuel tank is installed in an aircraft. Most tanks are constructed of noncorrosive material(s). They are typically made to be vented either through a vent cap or a vent line. Aircraft fuel tanks have a low area called a sump that is designed as a place for contaminants and water to settle. The sump is equipped with a drain valve used to remove the impurities during preflight walk-around inspection. Most aircraft fuel tanks contain some sort of baffling to subdue the fuel from shifting rapidly during flight maneuvers. Use of a scupper constructed around the fuel fill opening to drain away any spilled fuel is also common.

[6] On many aircraft, especially transport category and high-performance aircraft, part of the structure of the wings or fuselage is sealed with a fuel resistant two-part sealant to form a fuel tank. The sealed skin and structural members provide the highest volume of space available with the lowest weight. This type of tank is called an integral fuel tank since it forms a tank as a unit within the airframe structure (Fig. 10-1).

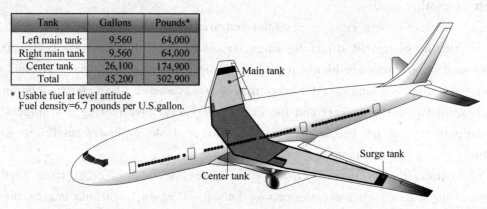

Fig. 10-1 Fuel tanks location

[7] When entering and performing maintenance on an integral fuel tank, all fuel must be emptied from the tank and strict safety procedures must be followed. Fuel vapors must be purged from the tank and respiratory equipment must be used by technicians. A full-time potter must be positioned just outside of the tank to assist if needed.

[8] **Fuel pump**: Other than aircraft with gravity-feed fuel systems, all aircraft have at least one fuel pump to deliver clean fuel under pressure to the fuel metering device for each engine. Engine-driven pumps are the primary delivery device. Auxiliary pumps are used on many aircraft as well. Sometimes known as booster pumps or boost pumps, auxiliary pumps are used to provide fuel under positive pressure to the engine-driven pump and during starting when the engine-driven pump is not yet up to speed for

sufficient fuel delivery. They are also used to back up the engine-driven pump during takeoff and at high altitude to guard against vapor lock. On many large aircraft, boost pumps are used to move fuel from one tank to another.

[9] The most common type of auxiliary fuel pump used on aircraft, especially large and high-performance aircraft, is the centrifugal pump. It is electric motor driven and most frequently submerged in the fuel tank or located just outside of the bottom of the tank with the inlet of the pump extending into the tank. If the pump is mounted outside the tank, a pump removal valve is typically installed so the pump can be removed without draining the fuel tank (Fig. 10-2).

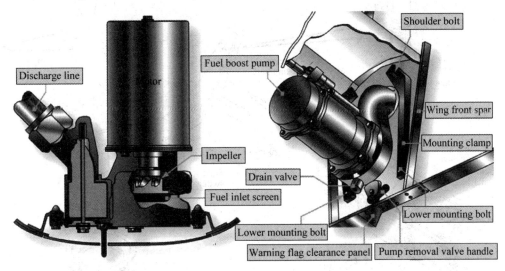

Fig. 10-2 Centrifugal fuel boost pump

[10] A centrifugal boost pump is a variable displacement pump. It takes in fuel at the center of an impeller and expels it to the outside as the impeller turns. An outlet check valve prevents fuel from flowing back through the pump. A fuel feed line is connected to the pump outlet. A bypass valve may be installed in the fuel feed system to allow the engine-driven pump to pull fuel from the tank if the boost pump is not operating. The centrifugal boost pump is used to supply the engine-driven fuel pump, back up the engine-driven fuel pump, and transfer fuel from tank to tank if the aircraft is so designed.

[11] Maintenance technicians are often asked to fuel or defuel aircraft. Fueling procedure can vary from aircraft to aircraft. Tanks may need to be fueled in a prescribed sequence to prevent structural damage to the airframe. The proper procedure should be confirmed before fueling an unfamiliar aircraft.

[12] **Fueling**: Refueling an aircraft is usually accomplished by a fuel truck. Refueling of all the tanks is done from the pressure fueling station. You can manually refuel the main tanks from the over-wing fueling ports. To manually refuel the center

tank, you transfer fuel from the main tanks to the center tank. In pressure fueling system, the fueling hose is coupled to the refueling receptacle located in the aircraft as shown in Fig. 10-3. When the hose or the hydrant is uncoupled, the receptacle will shut off and seal automatically. The refueling system can be controlled automatically by means of a refueling control panel which is controlled by the ground personnel. The fueling float switches prevent fueling the fuel tanks with too much fuel, they remove power to the fueling valves when the fuel level in the tank gets full.

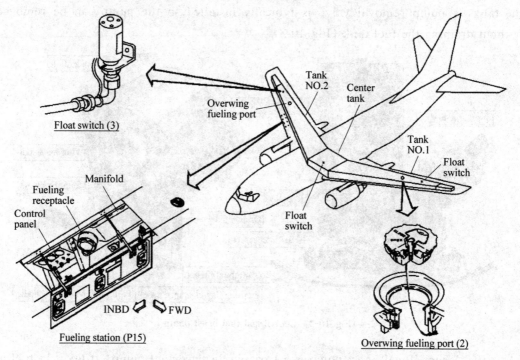

Fig. 10-3 A typical fueling system

[13] **Defueling**: Removing the fuel contained in aircraft fuel tanks is sometimes required. This can occur for maintenance, inspection, or due to contamination. Occasionally, a change in flight plan may require defueling. Safety procedures for defueling are the same as those for fueling: ①Always defuel outside; ②Fire extinguishers should be on hand; ③Bonding cables should be attached to guard against static electricity buildup; ④Defueling should be performed by experienced personnel, and inexperienced personnel must be checked out before doing so without assistance.

[14] Pressure fueled aircrafts are normally defuelled through the pressure fueling port. The aircraft's in-tank boost pumps can be used to pump the fuel out. The pump on a fuel truck can also be used to draw fuel out. These tanks can also be drained through the tank sump drains, but the large size of the tanks usually makes this impractical. Aircraft fueled over the wing are normally drained through the tank sump drains.

[15] A fuel jettison system is used to dump off enough fuel into air in case an

emergency landing is needed and the total weight of the aircraft is greater than its maximum landing weight. The maximum landing weight is a design specification of an aircraft which is used to limit the landing weight for protecting the aircraft structure from damage due to landing. A fuel jettison system must be able to jettison enough fuel within 10 minutes for general aviation, or 15minutes for transport category aircraft. It must be operable under the conditions encountered during all operations of the aircraft. The fuel jettison system is usually divided into two separate, independent systems, one for each wing, so that lateral stability can be maintained by jettisoning fuel from the "heavy" wing if it is necessary to do so. Normally, if an unbalanced fuel load exists fuel will be used from the "heavy" wing by supplying fuel to engines on the opposite wing. The system consists of valves, lines, dump chutes and chute-operating mechanisms. Each wing contains either a fixed or an extendable dump chute depending upon system design. In either case the fuel must be discharged clear of the airplane.

NEW WORDS AND EXPRESSIONS

(1) storage['stɔːrɪdʒ] 存储
(2) tank[tæŋk] 油箱
(3) ignition[ɪgˈnɪʃn] 点火
(4) lightning[ˈlaɪtnɪŋ] 闪电
(5) corona[kəˈrəʊnə] 电晕
(6) streamer[ˈstriːmər] 流化
(7) vent[vent] 通气
(8) luminous[ˈluːmɪnəs] 发光的；明亮的
(9) ionized[ˈaɪə,naɪz] 电离的
(10) imminent[ˈɪmɪnənt] 即将来临的；迫近的
(11) aviation[,eɪviˈeɪʃn] 航空
(12) gasoline[ˈɡæsəliːn; ,ɡæsəˈliːn] 汽油
(13) fractionally[ˈfrækʃənəlɪ] 微小地；极少地
(14) distill[dɪˈstɪl] 蒸馏
(15) volatility[,vɔləˈtɪlətɪ] 挥发性
(16) flashpoint[ˈflæʃpɔɪnt] 闪点
(17) freeze[ˈfriːz] 冻结
(18) wide-cut[ˈwaɪəkʌt] 宽馏分的
(19) kerosene[ˈkerəsiːn] 煤油
(20) refuel[,riːˈfjuːəl] 加油
(21) jettison[ˈdʒetɪsn] 应急放油
(22) bladder[ˈblædə(r)] 气囊，胶囊

(23) sump[sʌmp]　　　　　　　　　　集油槽
(24) subdue[səb'djuː]　　　　　　　　抑制；减轻
(25) scupper['skʌpə(r)]　　　　　　　排水口
(26) category['kætəgərɪ]　　　　　　　种类，分类
(27) sealant['siːlənt]　　　　　　　　密封剂
(28) skin[skɪn]　　　　　　　　　　　蒙皮
(29) centrifugal[ˌsentrɪ'fjuːgl]　　　　离心的
(30) boost[buːst]　　　　　　　　　　增压的，增加，提高
(31) expel[ɪk'spel]　　　　　　　　　喷出，排出
(32) maintenance['meɪntənəns]　　　　维护
(33) technician[teknɪʃən]　　　　　　　技术员
(34) defuel[ˌdiː'fjuːəl]　　　　　　　　放油
(35) nozzle['nɒzl]　　　　　　　　　　喷嘴

EXERCISE

1. Translate the followings into Chinese.

(1) **Fuel**. Any material that can be burned to release its energy. Gasoline, kerosine, coal, wood, and natural gas are all forms of fuel.

(2) **Lightning**. A bright flash of light in the atmosphere caused by a discharge of electricity from one cloud to another or between a cloud and the earth. Lightning is actually an extremely large spark.

(3) **Corona**. A prismatically colored circle or arc of a circle with the sun or moon at its center. The color of a corona changes from blue inside to red outside, which is opposite to that of a halo. Coronas vary in size and are much smaller than the fixed size of a halo.

(4) **Bladder-type fuel tank**. A plastic-impregnated fabric bag supported in a portion of an aircraft structure so that it forms a cell in which fuel is carried.

(5) **Flash point**. The temperature to which a material must be raised for it to ignite, but not continue to burn, when a flame is passed above it.

(6) **Cross-feed**: An interconnection between fuel manifolds for fuel flow from one side of the airplane to an adjacent engine or one on the other side.

(7) **Fuel jettison system**: The portion of an aircraft fuel system that allows fuel to be dumped from the tanks to lower the landing weight so a safe landing can be made. Some aircraft are certificated with a higher takeoff weight than is allowed for landing. If an emergency requires such an airplane to return for a landing before enough fuel is used from the tanks to lower the weight enough for landing, fuel may be dumped.

(8) **Fuel boost pump**. An auxiliary fuel pump used in an aircraft fuel system to provide fuel pressure for starting the engine, for use as an emergency backup in case the fuel pump fails, and to transfer fuel from one tank to another.

(9) **Dump chute**. A specially shaped duct, or tube, used to carry fuel away from the aircraft when it is dumped from the fuel tanks. Some large aircraft are permitted to take off with a greater weight than they are allowed to have when they land. However, if such an aircraft must make an emergency landing before it has burned off enough fuel to reduce its weight to its legal landing weight, it must dump enough fuel to get rid of this excess weight. The dump chute is designed to carry the fuel away from the aircraft so it will not be ignited by the engine exhaust or by static electricity.

(10) **Centrifugal pump**. A variable displacement pump which moves fluid by taking it into the center of a scroll-type impeller and slinging it outward by centrifugal action. Aircraft fuel boost pumps installed inside the fuel tanks are almost always centrifugal pumps.

2. Answer the following questions in English.
(1) What are the subsystems of aircraft fuel system?
(2) What is the sequence of the aircraft fuel feed?
(3) Why does the aircraft fuel tank need to be pressurized?
(4) What are the precautions for aircraft refueling?
(5) What is the function of a fuel jettison system on the aircraft?

SUPPLEMENTARY READING

[1] Fuel Leak Classification: Four basic classifications are used to describe aircraft fuel leaks: stain, seep, heavy seep, and running leak. In 30 minutes, the surface area of the collected fuel from a leak is a certain size. This is used as the classification standard. When the area is less than $\frac{3}{4}$ inch in diameter, the leak is said to be a stain. From $\frac{3}{4}$ to $1\frac{1}{2}$ inches in diameter, the leak is classified as a seep. Heavy seeps form an area from $1\frac{1}{2}$ inches to 4 inches in diameter. Running leaks pool and actually drip from the aircraft. They may follow the contour of the aircraft for a long distance.

[2] Two main types of fuel cleaning device are utilized on aircraft. Fuel strainers are usually constructed of relatively coarse wire mesh. They are designed to trap large pieces of debris and prevent their passage through the fuel system. Fuel strainers do not inhibit the flow of water. Fuel filters are usually fine mesh. In various applications, they can trap fine sediment that can be only thousands of an inch in diameter and also

help trap water. The technician should be aware that the terms "strainer" and "filter" are sometimes used interchange ably. Micronic filters are commonly used on turbine-powered aircraft. This is a type of filter that captures extremely fine particles in the range of 10-25 microns. A micronis1/1,000 of amillimeter.

[3] All aircraft fuel systems have filters and strainers to ensure that the fuel delivered to the engine(s) is free from contaminants. The first of these is encountered at the outlet of the fuel tank. A sump is used to encourage the collection of debris in the lowest part of the tank, which can then be drained off before flight. The actual tank outlet for the fuel is positioned above this sump. Some type of screen is used to trap contaminants attempting to flow out of the tank into the fuel system. Finger screens are common on light aircraft. They effectively increase the area of the fuel tank outlet, allowing a large amount of debris to be trapped while still permitting fuel to flow.

[4] All aircraft fuel systems must have some form of fuel quantity indicator. These devices vary widely depending on the complexity of the fuel system and the aircraft on which they are installed. Simple indicators requiring no electrical power were the earliest type of quantity indicators and are still in use today. The use of these direct reading indicators is possible only on light aircraft in which the fuel tanks are in close proximity to the cockpit. Other light aircraft and larger aircraft require electric indicators or electronic capacitance type indicators.

[5] A sight glass is a clear glass or plastic tube open to the fuel tank that fills with fuel to the same level as the fuel in the tank. It can be calibrated in gallons or fractions of a full tank that can be read by the pilot. Another type of sight gauge makes use of a float with an indicating rod attached to it. As the float moves up and down with the fuel level in the tank, the portion of the rod that extends through the fuel cap indicates the quantity of fuel in the tank. These two mechanisms are combined in yet another simple fuel quantity indicator in which the float is attached to a rod that moves up or down in a calibrated cylinder.

[6] More sophisticated mechanical fuel quantity gauges are common. A float that follows the fuel level remains the primary sensing element, but a mechanical linkage is connected to move a pointer across the dial face of an instrument. This can be done with a crank and pinion arrangement that drives the pointer with gears, or with a magnetic coupling, to the pointer.

[7] Electric fuel quantity indicators are more common than mechanical indicators in modern aircraft. Most of these units operate with direct current (DC) and use variable resistance in a circuit to drive a ratiometer-type indicator. The movement of a float in the tank moves a connecting arm to the wiper on a variable resistor in the tank unit. This resistor is wired in series with one of the coils of the ratiometer-type fuel gauge in the instrument panel. Changes to the current flowing through the tank unit resistor change the current flowing through hone of the coils in the indicator. This alters the

magnetic field in which the indicating pointer pivots. The calibrated dial indicates the corresponding fuel quantity.

[8] Digital indicators are available that work with the same variable resistance signal from the tank unit. They convert the variable resistance into a digital display in the cockpit instrument. Fully digital instrumentation systems, such as those found in a glass cockpit aircraft, convert the variable resistance into a digital signal to be processed in a computer and displayed on a flat screen panel.

TOPIC 10：ATA100 规范

在飞机运行和维修过程中会用到各种技术出版物，如表 T10-1 所示。这些技术出版物的编制应遵循一定的规范。

表 T10-1　技术出版物

缩写名称	英文全称及中文名称
AFM	Aircraft Flight Mauual，飞机飞行手册
AMM	Aircraft Maintenance Manual，飞机维护手册
ASM/SM/SSM	Aircraft System Schematics Manual，飞机系统图册
AWM/WM/WDM	Aircraft Wiring Diagram Manual，飞机线路图册
BCLM/CLS/CLM	Baggage/Cargo Loading System Manual，货物装载系统手册
BITE	Built-In Test Manual，自检手册
CMM	Component Maintenance Manual，部件维护手册
CPM	Corrosion Prevention Manual，防腐手册
DDG	Despatch Deviation Guide，放行偏离指南
FCOM	Flight Crew Operating Manual，机组操作手册
FIM	Fault Isolation Manual，故障隔离手册
FRM	Fault Reporting Manul，故障报告手册
IPC	Illustrated Parts Catalog，图解零部件目录
ITEM	Illustrated Tool And Equipment Manual，图解工具设备手册
MMEL	Main Minimum Equipment List，主最低设备清单
MPD	Maitenance Planning Data/Document，维修计划资料/文件
NDT/NTM	Non Destructive Testing Manual，无损检测手册
OHM	Overhaul Manual，大修手册
PPBM	Powerplant Buildup Manual，发动机总装手册
SRM	Structural Repair Manual，结构修理手册
SSM	System Schematic Manual，系统原理图手册
SWPM	Standard Wiring Practices Manual，标准线路施工手册

美国航空运输协会（Air Transport Association of America，ATA）规范是为了让制造业使用共同的系统和程序以达到降低成本的目的而发布的一系列规范。ATA 规范是对资料传递、处理业务、执行操作而发布的自发性的产业协议，其类别如表 T10-2 所示。

表 T10-2　ATA 规范

ATA 规范	内容
ATA Spec 100	Manufacturers Technical Data
ATA Spec 101	Ground Equipment Technical Data
ATA Spec 102	Computer Software Manual
ATA Spec 103	Standard for Jet Quality Control at Airports
ATA Spec 104	Guidelines for Aircraft Maintenance Training
ATA Spec 105	Guidelines for Training and Qualifying Personnel in Non-Destructive Testing Methods

续表

ATA 规范	内容
ATA Spec 106	Sources and Approved Parts Qualification Guidelines
ATA Spec 200	Integrated Data Processing Supply
ATA Spec 300	Packaging of Airline Supplies
ATA Spec 2000	Integrated Data Processing Materials Management
ATA Spec 2100	Digital Data Standards for Aircraft Support
ATA Spec 2200	Information Standards for Aviation Maintenance

其中，ATA100是航空产品技术资料编写规范，是美国航空运输协会会同航空产品制造商和航空公司共同制定的一种规范，用以统一各种民用航空产品制造商所出版的各种技术资料的编号。目前这一规范已经被很多国家所采用，成为一种民用航空器产品在设计、制造、使用和维修等各种资料、文件、函电、报告和目录索引的国际上的统一编号，使各种技术记录和数据处理趋于统一，改进了各种资料和文件的归档保管，促进民用航空各个领域的交流。

按照ATA章节，航空产品大体可以分为航空器和动力装置两大类，其中航空器又可分为：总体（General，5~12章）、系统（System，20~49章）、结构（Structure，51~57章），动力装置可分为螺旋/旋翼（Propeller/Rotor，60~65章）、发动机（Engine，70~91章），具体划分如表T10-3所示。

表 T10-3　ATA100 章节代码

ATA CODE LISTING		
5. time limits/maintenance. checks	28. fuel	53. fuselage
6. dimensions & areas	29. hydraulic power	54. nacelles/pylons
7. lifting & shoring	30. ice & rain protection	55. stabilizers
8. leveling & weighing	31. indicating/recording systems	56. windows
9. towing & taxiing	32. landing gear	57. wings
10. parking & mooring	33. lights	70. standard practices engine
11. placards &markings	34. navigation	71. power plant
12. servicing	35. oxygen	72. engine
20. standard practices airframe	36. pneumatic	73. engine fuel & control
21. air conditioning	37. vacuum	74. ignition
22. auto flight	38. water/waste	75. air
23. communications	45. central maintenance system	76. engine controls
24. electrical power	47. inerting gas system	77. engine indicating
25. equipment/furnishings	49. APU	78. exhaust
26. fire protection	51. standard practices structures	79. oil
27. flight controls	52. doors	80. starting

上表中每一章对应的就是飞机上的一个系统，是由各相关零件组成的能完成某种特定功能的集合。章的编号作为第一组号码；子系统是系统的某一部分，每一个系统可以由几个子系统组成，子系统由节号指定一个编号作为第二组号码；单件是组成系统或子系统的各零部件，以及单独的管路、线路等，单件的编号作为第三组号码，由零部件制造商自行编

排,示例如图 T10-1 所示。

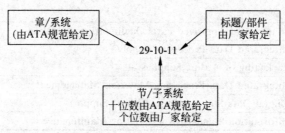

图 T10-1　ATA 编号规则

为了能迅速找到手册中的相关内容,ATA100 对维护手册、翻修手册的页码范围按不同的内容做出了规定,如表 T10-4 所示。

表 T10-4　手册页码

页 码 范 围	内容(维护手册)	内容(翻修手册)
1~99	说明及工作	说明及工作
101~199	故障分析	分解
201~299	维护实施	清洗
301~399	勤务	检验/检查
401~499	拆卸/安装	修理
501~599	调节/测试	装配
601~699	检验/检查	间隙容差
701~799	清洁/喷漆	试验
801~899	核准修理方法	故障分析
901~999		存放
1001~1099		特种工具、夹具及设备
1101~1199		图解零件目录
1201~1299		小翻修

UNIT 11

HYDRAULIC SYSTEM

[1] The word hydraulics is based on the Greek word for water, and originally meant the study of the physical behavior of water at rest and in motion. Today the meaning has been expanded to include the physical behavior of all liquids, including hydraulic fluid.

[2] There are multiple applications for hydraulic use in aircraft, depending on the complexity of the aircraft. For example, hydraulics is often used on small airplanes to operate wheel brakes, retractable landing gear, and some constant-speed propellers. On large airplanes, hydraulics is used for flight control surfaces, wing flaps, spoilers, and other systems. A basic hydraulic system consists of a reservoir, pump (either hand, electric, or engine driven), a filter to keep the fluid clean, selector valve to control the direction of flow, relief valve to relieve excess pressure, and an actuator (Fig. 11-1).

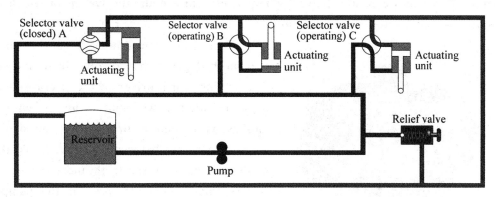

Fig. 11-1 Basic hydraulic system

[3] The hydraulic fluid is pumped through the system to an actuator or servo. A servo is a cylinder with a piston inside that turns fluid power into work and creates the power needed to move an aircraft system or flight control. Servos can be either single-acting or double-acting, based on the needs of the system. This means that the fluid can be applied to one or both sides of the servo, depending on the servo type. A single-acting servo provides power in one direction. The selector valve allows the fluid direction to be controlled. This is necessary for operations such as the extension and retraction of landing gear during which the fluid must work in two different directions. The relief valve provides an outlet for the system in the event of excessive fluid pressure

in the system. Each system incorporates different components to meet the individual needs of different aircraft.

[4] The aircraft has three main hydraulic systems. They are identified as the Green, Blue and Yellow systems (A, B, C systems identified on some aircrafts). Together they supply hydraulic power at 3,000 psi (5,000psi for A380 and B787) to the main power users. These include the flight controls, landing gear, cargo doors, brakes and thrust reversers. Services which are not used during flight (cargo doors, brakes, landing gear and nosewheel steering) are isolated from the main supply.

[5] **Power element**: The function of the power element is to transform the mechanical energy of the prime mover into the pressure energy of the liquid, which means the pump in the hydraulic system, and it provides power to the whole hydraulic system. The types of hydraulic pump are generally gear pump, vane pump, piston pump and screw pump.

[6] A constant delivery pump, regardless of pump rpm, forces a fixed or unvarying quantity of fluid through the outlet port during each revolution of the pump. Constant delivery pumps are sometimes called constant volume or fixed delivery pumps. They deliver a fixed quantity of fluid per revolution, regardless of the pressure demands. Since the constant delivery pump provides a fixed quantity of fluid during each revolution of the pump, the quantity of fluid delivered per minute will depend upon pump rpm. When a constant delivery pump is used in a hydraulic system in which the pressure must be kept at a constant value, a pressure regulator is required.

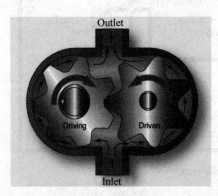

Fig. 11-2　Gear pump

[7] A gear type power pump consists of two meshed gears that revolve in a housing. The driving gear is driven by the aircraft engine or some other power unit. The driven gear meshes with, and is driven by, the driving gear. Clearance between the teeth as they mesh, and between the teeth and the housing, is very small. The inlet port of the pump is connected to the reservoir, and the outlet port is connected to the pressure line. When the driving gear turns in a clockwise direction, as shown in Fig.11-2, it turns the driven gear in a counterclockwise direction. As the gear teeth pass the inlet port, fluid is trapped between the gear teeth and the housing, and is then carried around the housing to the outlet port.

[8] A variable delivery pump has a fluid output that is varied to meet the pressure demands of the system by varying its fluid output. The pump output is changed automatically by a pump compensator within the pump. Fig.11-3 shows an axial-piston, variable-displacement, yoke-actuated, pressure compensated hydraulic pump with a

solenoid-operated depressurization valve.

[9] The pump has a rotating cylinder barrel that has nine pistons. The pistons ride on an inclined yoke surface. As the barrel rotates, the pistons reciprocate within the barrel. Change to the incline of the yoke changes the stroke of each piston and changes the variable-displacement of the pump. During the intake stroke of each piston, hydraulic fluid is drawn in the top of the barrel and into the piston cylinder bores. During the discharge stroke of each piston, hydraulic fluid is forced out of the piston cylinder bores and into the output pressure line. Some of the supply fluid in the pump becomes case drain fluid. The case drain fluid cools and lubricates the pump before it goes to the heat exchanger and reservoir through the case drain connection. A pressure compensator in the pump maintains a preset pressure by control of the fluid flow out of the pump in response to changes in system demands. The pump operates in the pressurized or depressurized mode. In the depressurized mode, the pump is isolated from the hydraulic system and runs at a zero output flow. The solenoid operated depressurization valve controls the pressurized or depressurized modes of the pump.

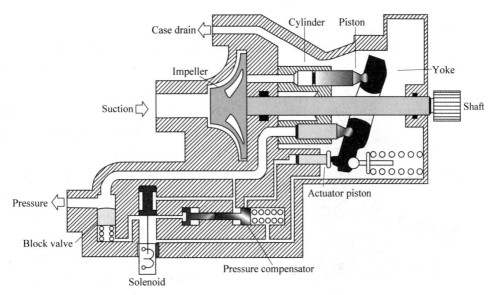

Fig. 11-3 Variable delivery pump

[10] **Control element**: The function of control elements is to control and regulate the pressure, flow and direction of liquid in the hydraulic system. According to the different control functions, the hydraulic valve can be divided into pressure control valve, flow control valve and direction control valve.

[11] A selector valve is used to control the direction of movement of a hydraulic actuating cylinder or similar device (Fig. 11-4). It provides for the simultaneous flow of hydraulic fluid both into and out of the unit. Hydraulic system pressure can be routed with the selector valve to operate the unit in either direction and a corresponding return

path for the fluid to the reservoir is provided.

[12] A check valve allows fluid to flow unimpeded in one direction but prevents or restricts fluid flow in the opposite direction (Fig. 11-5). A check valve may be an independent component situated in-line somewhere in the hydraulic system or it may be built-in to a component. When part of a component, the check valve is said to be an integral check valve.

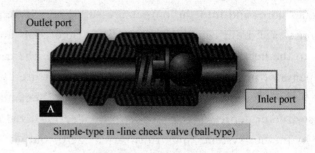

Fig. 11-4　Selector valve　　　　　Fig. 11-5　Check valve

[13] A pressure relief valve is used to limit the amount of pressure being exerted on a confined liquid. This is necessary to prevent failure of components or rupture of hydraulic lines under excessive pressures. The pressure relief valve is, in effect, a system safety valve. The design of pressure relief valves incorporates adjustable spring-loaded valves. They are installed in such a manner as to discharge fluid from the pressure line into a reservoir return line when the pressure exceeds the predetermined maximum for which the valve is adjusted. Various makes and designs of pressure relief valves are in use, but, in general, they all employ a spring-loaded valving device operated by hydraulic pressure and spring tension. The two-port valve is illustrated in Fig. 11-6.

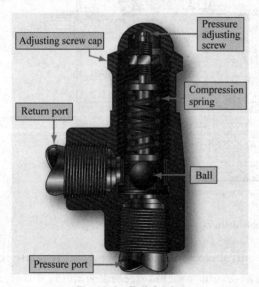

Fig. 11-6　Relief valve

[14] **Actuate element**: An actuating cylinder transforms energy in the form of fluid pressure into mechanical force, or action, to perform work. It is used to impart powered linear motion to some movable object or mechanism. A typical actuating cylinder consists of a cylinder housing, one or more pistons and piston rods, and some seals. The cylinder housing contains a polished bore in which the piston operates, and one or more ports through which fluid enters and leaves the bore. The piston and rod

form an assembly. The piston moves forward and backward within the cylinder bore, and an attached piston rod moves into and out of the cylinder housing through an opening in one end of the cylinder housing (Fig. 11-7).

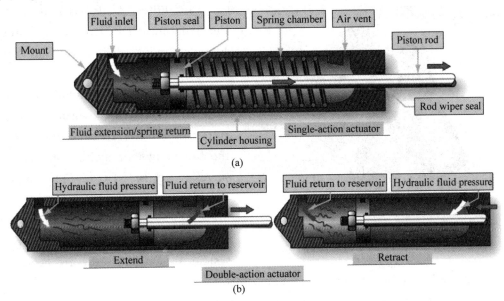

Fig. 11-7 Actuator

[15] Actuating cylinders are of two major types: single-action and double-action. The single-action (single port) actuating cylinder is capable of producing powered movement in one direction only. The double-action (two ports) actuating cylinder is capable of producing powered movement in two directions.

[16] **Auxiliary element**: includes pressure gauge, filter, accumulator, heat exchanger, tubing and reservoir, etc. Only accumulator, filter, and reservoir will be introduced here.

[17] **Accumulator**: The accumulator is a steel sphere divided into two chambers by a synthetic rubber diaphragm. The upper chamber contains fluid at system pressure, while the lower chamber is charged with nitrogen or air. Cylindrical types are also used in high-pressure hydraulic systems. Many aircraft have several accumulators in the hydraulic system. There may be a main system accumulator and an emergency system accumulator. There may also be auxiliary accumulators located in various sub-systems. The function of an accumulator is to:
- Dampen pressure surges in the hydraulic system caused by actuation of a unit and the effort of the pump to maintain pressure at a preset level.
- Aid or supplement the power pump when several units are operating at once by supplying extra power from its accumulated, or stored, power.
- Store power for the limited operation of a hydraulic unit when the pump is not operating.

- Supply fluid under pressure to compensate for small internal or external (not desired) leaks that would cause the system to cycle continuously by action of the pressure switches continually kicking in.

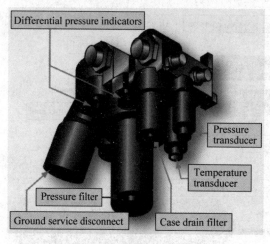

Fig. 11-8　Filter

[18] **Filter**: A filter is used to clean the hydraulic fluid, preventing foreign particles and contaminating substances from remaining in the system (Fig. 11-8). If such objectionable material were not removed, the entire hydraulic system of the aircraft could fail through the breakdown or malfunctioning of a single unit of the system. Filters may be located within the reservoir, in the pressure line, in the return line, or in any other location the designer of the system decides that they are needed to safeguard the hydraulic system against impurities.

[19] The filter assembly is comprised of three basic units: head assembly, bowl, and element. The head assembly is secured to the aircraft structure and connecting lines. Within the head, there is a bypass valve that routes the hydraulic fluid directly from the inlet to the outlet port if the filter element becomes clogged with foreign matter. The bowl is the housing that holds the element to the filter head and is removed when element removal is required. The element may be a micron, porous metal, or magnetic type. The micron element is made of a specially treated paper and is normally thrown away when removed. The porous metal and magnetic filter elements are designed to be cleaned by various methods and replaced in the system.

[20] **Reservoir**: The reservoir is a tank in which an adequate supply of fluid for the system is stored. Fluid flows from the reservoir to the pump, where it is forced through the system and eventually returned to the reservoir. The reservoir not only supplies the operating needs of the system, but it also replenishes fluid lost through leakage. Furthermore, the reservoir serves as an overflow basin for excess fluid forced out of the system by thermal expansion (the increase of fluid volume caused by temperature changes), the accumulators, and by piston and rod displacement.

[21] The reservoir also furnishes a place for the fluid to purge itself of air bubbles that may enter the system. Foreign matter picked up in the system may also be separated from the fluid in the reservoir or as it flows through line filters. Baffles and/or fins are incorporated in most reservoirs to keep the fluid within the reservoir from having random movement, such as vortexing (swirling) and surging. These conditions can cause fluid to foam and air to enter the pump along with the fluid. Many reservoirs

incorporate strainers in the filler neck to prevent the entry of foreign matter during servicing.

[22] Air-pressurized reservoirs are used in many commercial transport-type aircraft (Fig. 11-9). Pressurization of the reservoir is required because the reservoirs are often located in wheel wells or other non-pressurized areas of the aircraft and at high altitude there is not enough atmospheric pressure to move the fluid to the pump inlet. Engine bleed air is used to pressurize the reservoir. The reservoirs are typically cylindrical in shape. The following components are installed on a typical reservoir: pressure relief valve, sight glasses, sample valve, drain valve, temperature transducer, quantity transmitter. A pressurization module is installed close to the reservoir. The pressurization module supplies airplane bleed air to the reservoirs.

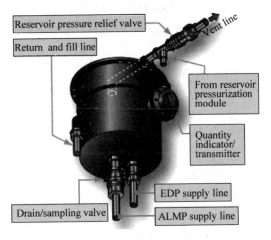

Fig. 11-9 Air-pressurized reservoir

NEW WORDS AND EXPRESS

(1) propeller[prə'pelər]　　　　螺旋桨
(2) reservoir['rezəvwɑː]　　　　液压油箱
(3) pump[pʌmp]　　　　　　　泵
(4) filter['fɪltə]　　　　　　　过滤器,油滤
(5) valve[vælv]　　　　　　　阀,活门
(6) actuator['æktʃueɪtə]　　　　执行机构;作动器
(7) servo['sɜːvəʊ]　　　　　　伺服;伺服系统
(8) cylinder['sɪlɪndə]　　　　　作动筒;柱塞,瓶体
(9) thrust reverser　　　　　　反推装置
(10) power element　　　　　　动力元件
(11) mechanical energy　　　　机械能
(12) regulator['regjʊleɪtə]　　　调节器
(13) housing['haʊzɪŋ]　　　　　外壳;机匣
(14) clearance['klɪərəns]　　　间隙;清除;许可
(15) cylinder barrel　　　　　　柱塞缸体
(16) yoke[jəʊk]　　　　　　　斜盘
(17) piston['pɪstən]　　　　　　活塞

(18) stroke[strəʊk] 行程,冲程
(19) solenoid['sɔlənɔɪd; 'səʊlənɔɪd] 电磁线圈
(20) unimpeded[ˌʌnɪm'piːdɪd] 畅通无阻的；未受阻的
(21) rupture['rʌptʃə(r)] 破裂,决裂；使破裂
(22) polish['pɔlɪʃ] 抛光,擦亮；磨光
(23) gauge[geɪdʒ] 计量器,表
(24) accumulator[ə'kjuːmjəleɪtə(r)] 蓄压器
(25) sphere[sfɪə(r)] 球体的
(26) chamber['tʃeɪmbə(r)] 腔,室
(27) synthetic[sɪn'θetɪk] 合成物；综合的,合成的,人造的
(28) nitrogen['naɪtrədʒən] 氮气
(29) impurity[ɪm'pjʊərɪtɪ] 杂质
(30) bypass valve 旁通活门
(31) micron['maɪkrɔn] 微米
(32) leakage['liːkɪdʒ] 泄漏；渗漏物；漏出量
(33) thermal['θəːml] 热力的
(34) volume['vɔljuːm] 体积
(35) bubble['bʌbl] 气泡,泡沫
(36) baffle['bæfl] 挡板
(37) vortex['vɔːteks] 涡流；漩涡
(38) strainer['streɪnə(r)] 过滤器

EXERCISE

1. **Translate the followings into Chinese.**
(1) **Reservoir**. The reservoir is a tank in which an adequate supply of fluid for the system is stored. Many reservoirs are equipped with a dipstick or a glass sight gauge by which fluid level can be conveniently and accurately checked.
(2) **Actuating cylinder**. A cylinder and piston arrangement used to convert hydraulic or pneumatic fluid pressure into work. Fluid under pressure moves the piston that does the work.
(3) **Accumulator**. A component in a hydraulic system that allows anon compressible fluid, such as oil, to be stored under pressure. An accumulator has two compartments separated by a flexible or movable partition such as a diaphragm, bladder, or piston. One compartment contains compressed air or nitrogen, and the other is connected into the source of hydraulic pressure.
(4) **Bypass valve**. A valve used to maintain a specific pressure in a fluid power system by bypassing some of the fluid back to the inlet of the system-pressure pump.

(5) **Gage**. A gage (also spelled gauge) is a measuring instrument. There are many different types of gages. A pressure gage is used to indicate the amount of pressure being applied to a fluid. A depth gage is used to measure the distance between the edge of a hole or groove and its bottom. A thickness gage is used to measure the clearance between close-fitting parts of a machine.

(6) **Filter**. A component in a system that traps and holds contaminants carried in the fluid flowing through the system. Aircraft fuel systems, lubrication systems, and hydraulic systems all use filters.

(7) **Check valve**. The valve allows free flow in one direction only, it prevents flow in the reverse direction.

(8) **Pressure relief valve**. A valve which prevents excessive pressure in a system by venting the system to ambient.

(9) **Variable-Displacement Pump**. A variable-displacement pump has a fluid output that is varied to meet the pressure demands of the system. The pump output is changed automatically by a pump compensator within the pump.

(10) **Servo**. A component in an automatic flight control system that actually moves the flight control. There are servos in each of the primary controls: the ailerons, elevators, and rudder. The automatic pilot senses when a flight correction is needed, and it sends a signal to the servo to move the control surface in the proper direction to make the correction.

2. Answer the following questions in English.

(1) Which elements are included in a basic hydraulic system?
(2) What is the function of the select valve in a hydraulic system?
(3) What is the function of the relief valve in a hydraulic system?
(4) What is the function of the power element (pump) in a hydraulic system?
(5) What is the function of the actuate element (actuator) in a hydraulic system?
(6) What are the functions of the accumulator in a hydraulic system?
(7) What is the function of the filter in a hydraulic system?
(8) What are the functions of the reservoir in a hydraulic system?

SUPPLEMENTARY READING

[1] To assure proper system operation and to avoid damage to nonmetallic components of the hydraulic system, the correct fluid must be used. When adding fluid to a system, use the type specified in the aircraft manufacturer's maintenance manual or on the instruction plate affixed to the reservoir or unit being serviced. There are three types of hydraulic fluids currently being used in civil aircraft.

[2] **Polyalphaolefin-based fluids**: MIL-H-83282 is a fire-resistant hydrogenated

polyalphaolefin-based fluid developed in the 1960s to overcome the flammability characteristics of MIL-H-5606. MIL-H-83282 is significantly more flame resistant than MIL-H-5606, but a disadvantage is the high viscosity at low temperature. It is generally limited to -40°F. However, it can be used in the same system and with the same seals, gaskets, and hoses as MIL-H-5606. MIL-H-46170 is the rust-inhibited version of MIL-H-83282. Small aircraft predominantly use MIL-H-5606, but some have switched to MIL-H-83282 if they can accommodate the high viscosity at low temperature.

[3] **Mineral-based fluids.** Mineral oil-based hydraulic fluid (MIL-H-5606) is the oldest, dating back to the 1940s. It is used in many systems, especially where the fire hazard is comparatively low. MIL-H-6083 is simply a rust-inhibited version of MIL-H-5606. They are completely interchangeable. Suppliers generally ship hydraulic components with MIL-H-6083. Mineral-based hydraulic fluid (MIL-H-5606) is processed from petroleum. It has an odor similar to penetrating oil and is dyed red. Some synthetic hydraulic fluids are dyed purple and even green, depending on the identity of the fluid. Synthetic rubber seals are used with petroleum-based fluids.

[4] **Phosphate ester base fluids.** These fluids are used in most commercial transport category aircraft and are extremely fire-resistant. However, they are not fireproof and under certain conditions, they burn. In addition, these fluids are very susceptible to contamination from water in the atmosphere. The earliest generation of these fluids was developed after World War II as a result of the growing number of aircraft hydraulic brake fires that drew the collective concern of the commercial aviation industry. Progressive development of these fluids occurred as a result of performance requirements of newer aircraft designs. The airframe manufacturers dubbed these new generations of hydraulic fluid, such as Skydrol and Hyjet, as types based on their performance.

[5] Several types of phosphate ester base (Skydrol) hydraulic fluids have been discontinued. Currently used in aircraft are Skydrol 500B-a clear purple liquid having good low temperature operating characteristics and low corrosive side effects; and, Skydrol LD-a clear purple low weight fluid formulated for use in large and jumbo jet transport aircraft where weight is a prime factor.

[6] Do not mix hydraulic fluid of the phosphate ester type with hydraulic fluid of the mineral-base type. This mixture causes a jelly in the hydraulic systems which can damage the system. This could cause are duction in aircraft safety. These types of fluid are used in the hydraulic systems and landing gear respectively. Thus, there must be precautions to prevent any mixture of the fluids. Make sure that the hydraulic fluid does not touch the aircraft other than components in the hydraulic system. Keep to a minimum the hydraulic fluid which falls accidentally during maintenance. Clean up such hydraulic fluid so that it can not go into adjacent areas. This is also to prevent future incorrect reports of hydraulic fluid leaks. When you clean up hydraulic fluid, use dry cloth and clean the area with cleaning agents.

TOPIC 11：飞机站位及区域划分

1. 飞机站位

站位这个概念源自制造业中的工位。在飞机制造的过程中，各个部件是由不同车间完成的，最后进行组装。不同的车间有不同的工位，因此，不同的部件也应有不同的站位标准。早期的小飞机曾把机头定义为 0 点，但是，如果后期在机头前加装部件，就会出现负数站位。随着飞机的逐渐变大，制造业分工也更为明确。在机头前留有足够空间，不仅符合工位的实际情况，也便于修正后期的构型改变，因此，站位线 station 就是工位线。

描述空间内的某一点，单有一条线远远不够，真实的世界需要三个维度。波音和空客都离不开这个思想，但也略有不同。谈起波音和空客的差异，很多人都会提到"人机关系"：波音飞机倾向于人的主观性，空客飞机更偏爱机器的精密度。从对定位的描述也能看出这点，波音飞机对坐标的定义角度更加通俗，而空客飞机直接使用了 *XYZ* 三轴。对于单位的选择，在科学领域，美国人更喜欢英制，欧洲人更喜欢公制，站位描述沿用了这一特点。下面以波音飞机为例进行说明，不同飞机稍有差异，具体可参看维护手册中相应章节。

1）机体基准尺寸

飞机上部件的位置可以根据标注尺寸来进行定位，如图 T11-1 所示，包括：

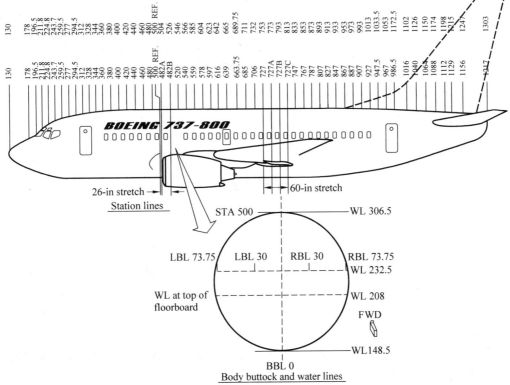

图 T11-1　机体站位

机身站位线(station, STA): 是水平标注尺寸,始于零站位线,从飞机前部的垂直基准面测量机身站位线。

机身纵剖线(buttock line, BL): 是横向标注尺寸,可测量机身中线向左或向右的纵剖线。

水线(waterline, WL): 是垂直标注尺寸,从飞机下方的水平基准面测量水线。

2) 机翼基准尺寸

机翼有两个基准尺寸(图 T11-2),以英寸的形式给出机翼部件位置,从机身纵剖线 0 开始测量,包括:

机翼站位(wing station, WS): 垂直于机翼前缘测量。

机翼纵剖线(wing buttock line, WBL): 平行于机身纵剖线测量。

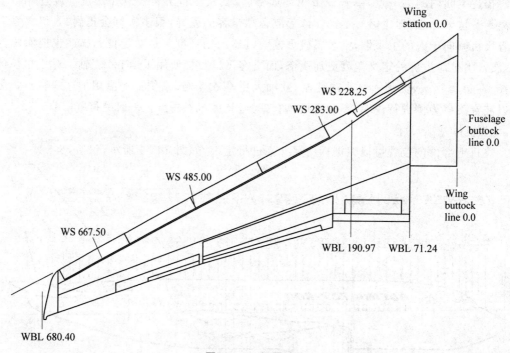

图 T11-2 机翼站位

2. 安定面站位

1) 垂直安定面基准尺寸

垂直安定面有四个基本尺寸,如图 T11-3 所示,以英寸为单位给出垂直安定面的位置。

垂直安定面站位(vertical stabilizer station): 垂直于垂直安定面后梁测量,零站位起始于顶线。

垂直安定面前缘站位(vertical stabilizer leading edge station): 垂直于垂直安定面前缘测量,零站位起始于顶线。

方向舵站位(rudder station): 垂直于方向舵铰链中心测量,零站位起始于顶线。

垂直安定面水线(vertical stabilizer waterline): 平行于机身水线测量。

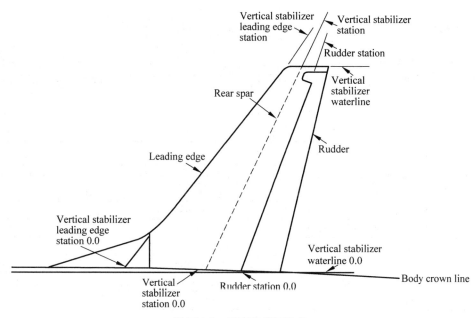

图 T11-3 垂直安定面站位

2）水平安定面基准尺寸

水平安定面有三个基准尺寸，以英寸为单位给出水平安定面位置，所有位置都是从机身纵剖线 0 开始测量，如图 T11-4 所示。

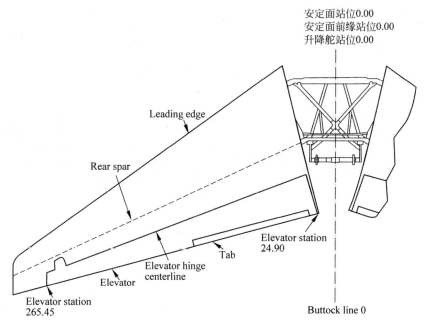

图 T11-4 水平安定面站位

水平安定面站位（horizontal stabilizer station）：垂直于水平安定面后梁测量。

水平安定面前缘站位（horizontal stabilizer leading edge station）：垂直于水平安定面

前缘测量。

升降舵站位（rudder station）：垂直于升降舵铰链中心测量。

3. 飞机区域划分

为了维护方便，飞机可以划分成不同的区域，不同飞机稍有不同，如表 T11-1 和图 T11-5、图 T11-6 所示。

表 T11-1　飞机区域划分

Zone	B737NG	A320
100	Lower half of fuselage	Lower half of fuselage to aft pressure bulkhead
200	Upper half of fuselage	Upper half of fuselage to aft pressure bulkhead
300	Empennage	Stabilizers
400	Power plant and nacelle struts	Nacelles
500	Left wing	Left wing
600	Right wing	Right wing
700	Landing gear and landing gear doors	Landing gear
800	Doors	Doors

具体可以参考相应机型 AMM 手册中的 ATA06 章。

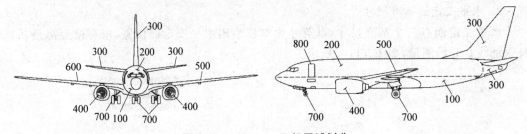

图 T11-5　B737NG 飞机区域划分

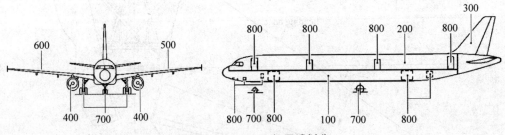

图 T11-6　A320 飞机区域划分

UNIT 12

LANDING GEAR SYSTEM

[1] Aircraft landing gear supports the entire weight of an aircraft during landing and ground operations. They are attached to primary structural members of the aircraft. The type of gear depends on the aircraft design and its intended use. Most landing gear has wheels to facilitate operation to and from hard surfaces, such as airport runways. Aircraft that operates to and from frozen lakes and snowy areas may be equipped with landing gear that have skis. Aircraft that operates to and from the surface of water have pontoon-type landing gear. Regardless of the type of landing gear utilized, shock absorbing equipment, brakes, retraction mechanisms, controls, warning devices, cowling, fairings, and structural members necessary to attach the gear to the aircraft are considered parts of the landing gear system. Three basic arrangements of landing gear are used: tail wheel-type landing gear (also known as conventional gear), tricycle-type landing gear, and tandem landing gear.

[2] **Tail wheel-type landing gear**: A tail wheel-type landing gear is also known as conventional gear because many early aircraft use this type of arrangement. The main gear is located forward of the center of gravity, causing the tail to require support from a third wheel assembly. A few early aircraft designs use a skid rather than a tail wheel. This helps slow the aircraft upon landing and provides directional stability. The resulting angle of the aircraft fuselage, when fitted with conventional gear, allows the use of a long propeller that compensates for older, underpowered engine design. The increased clearance of the forward fuselage offered by tail wheel-type landing gear is also advantageous when operating in and out of non-paved runways. Today, aircraft are manufactured with conventional gear for this reason and for the weight savings accompanying the relatively light tail wheel assembly.

[3] **Tandem landing gear**: Few aircraft are designed with tandem landing gear. As the name implies, this type of landing gear has the main gear and tail gear aligned on the longitudinal axis of the aircraft. Sailplanes commonly use tandem gear, although many only have one actual gear forward on the fuselage with a skid under the tail. A few military bombers, such as the B-47 and the B-52, have tandem gear, as does the U2 spy plane. The VTOL Harrier has tandem gear but uses small outrigger gear under the wings

for support. Generally, placing the gear only under the fuselage facilitates the use of very flexible wings.

[4] **Tricycle-type landing gear**: The most commonly used landing gear arrangement is the tricycle-type landing gear. It is comprised of main gear and nose gear. Nearly all aircraft have steerable nose gear. On light aircraft, the nose gear is directed through mechanical linkage to the rudder pedals. Heavy aircraft typically utilizes hydraulic power to steer the nose gear. Control is achieved through an independent tiller in the flight deck. The main gear on a tricycle-type landing gear arrangement is attached to reinforced wing structure or fuselage structure.

[5] The number and location of wheels on the main gear vary. Many main gears have two or more wheels. Multiple wheels spread the weight of the aircraft over a larger area. They also provide a safety margin should one tire fail. Heavy aircraft may use four or more wheels assemblies on each main gear. When more than two wheels are attached to a landing gear strut, the attaching mechanism is known as a bogie. The number of wheels included in the bogie is a function of the gross design weight of the aircraft and the surface type on which the loaded aircraft is required to land. The tricycle-type landing gear arrangement consists of many parts and assemblies. These include air/oil shock struts, gear alignment units, support units, retraction and safety devices, steering systems, wheel and brake assemblies, etc. A main landing gear of a transport category aircraft is illustrated in Fig.12-1.

[6] **Large aircraft retraction systems**: Large aircraft retraction systems are nearly always powered by hydraulics. Typically, the hydraulic pump is driven by the engine accessory drive. Auxiliary electric hydraulic pumps are also common. Other devices used in a hydraulically-operated retraction system include actuating cylinders, selector valves, uplocks, downlocks, sequence valves, tubing, and other conventional hydraulic system components. These units are interconnected so that they permit properly sequenced retraction and extension of the landing gear and the landing gear doors.

[7] The correct operation of any aircraft landing gear retraction system is extremely important. Fig.12-2 illustrates an example of a simple large aircraft hydraulic landing gear system. The system is on an aircraft that has doors that open before the gear is extended and close after the gear is retracted. The nose gear doors operate via mechanical linkage and do not require hydraulic power. There are many gear and gear door arrangements on various aircraft. Some aircrafts have gear doors that close to fair the wheel well after the gear is extended. Others have doors mechanically attached to the outside of the gear so that when it stows inward, the doors stow with the gear and fair with the fuselage skin.

[8] When the flight deck gear select lever is moved to the gear-up position, it positions a select valve to allow hydraulic pump pressure from the hydraulic system manifold to access eight different components. The three downlocks are pressurized and

UNIT 12 LANDING GEAR SYSTEM 143

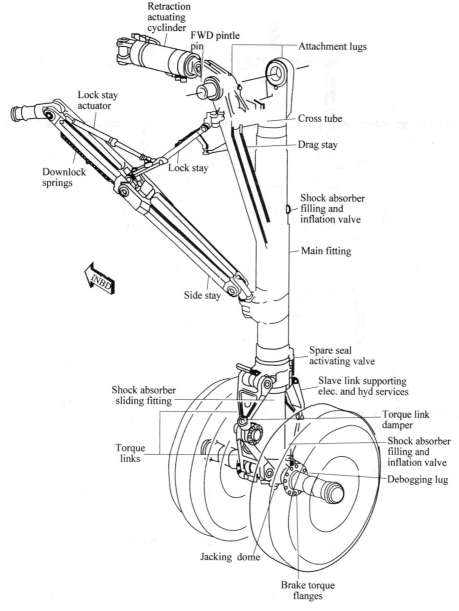

Fig. 12-1 Main landing gear

unlocked so the gear can be retracted. At the same time, the actuator cylinder on each gear also receives pressurized fluid to the gear-up side of the piston through an unrestricted orifice check valve. This drives the gear into the wheel well. Two sequence valves (C and D) also receive fluid pressure. Gear door operation must be controlled so that it occurs after the gear is stowed. The sequence valves are closed and delay flow to the door actuators. When the gear cylinders are fully retracted, they mechanically contact the sequence valve plungers that open the valves and allow fluid to flow into the close side of the door actuator cylinders. This closes the doors. Sequence valves A and B

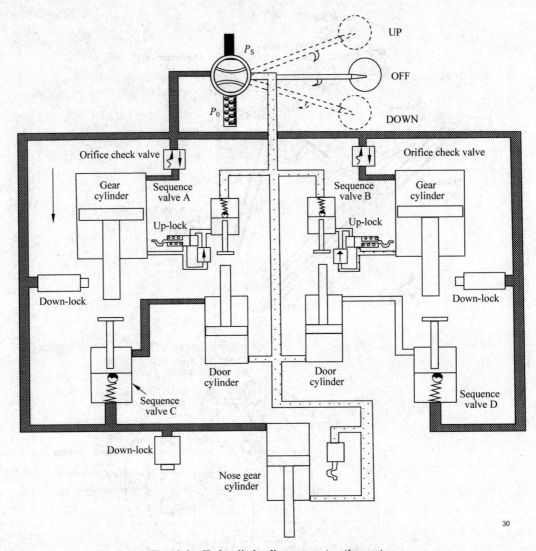

Fig. 12-2　Hydraulic landing gear retraction system

act as check valves during retraction. They allow fluid to flow one way from the gear-down side of the main gear cylinders back into the hydraulic system return manifold through the select valve.

[9] The emergency extension system lowers the landing gear if the main power system fails. There are numerous ways in which is done depending on the size and complexity of the aircraft. Some airrafts have emergency release handles in the flight deck that are connected through mechanical linkages to the gear uplocks. When the handle is operated, it releases the uplocks and allows the gear to free-fall to the extended position under the force created by gravity acting upon the gear. Other airrafts use a non-mechanical back-up, such as pneumatic power, to unlatch the gear.

[10] **Nose wheel steering systems**: The nose wheel on most aircraft is steerable from

the flight deck via a nose wheel steering system. This allows the aircraft to be directed during ground operation. A few simple aircraft are steered during taxi by differential braking. The nosewheel steering wheel connects through a shaft to a steering drum located inside the cockpit control pedestal. The rotation of this drum transmits the steering signal by means of cables and pulleys to the control drum of the differential assembly. Movement of the differential assembly is transmitted by the differential link to the metering valve assembly, where it moves the metering valve to the selected position. Then hydraulic pressure provides the power for turning the nose gear.

[11] As shown in Fig. 12-3, pressure from the aircraft hydraulic system is directed through the opened safety shutoff valve and into a line leading to the metering valve. This metering valve then routes pressurized fluid out of port A, through the right-turn alternating line, and into lower chamber of steering cylinder A and upper chamber of steering cylinder B. The pressure forces the cylinder A piston to start extending and cylinder B piston to start retracting. Since the rod of this two pistons connects to the nose steering spindle, which pivots at point X, the movement of this two pistons turns the steering spindle gradually toward the right. This action turns the nose gear slowly toward the right, because the spindle is connected to the nose gear shock strut. As the nose gear turns right, fluid is forced out of lower chamber of cylinder B and upper chamber of cylinder A, through the left turn alternating line, and into port B of the metering valve. The metering valve sends this return fluid into a compensator, which routes the fluid into the aircraft system return manifold.

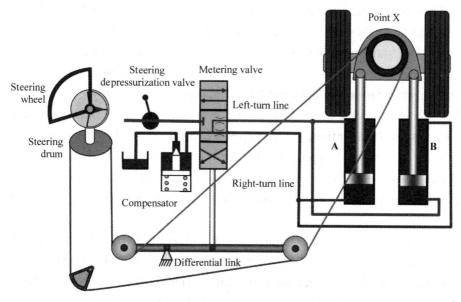

Fig. 12-3 Nosewheel steering system

[12] **Aircraft wheels**: Aircraft wheels are an important component of a landing gear system. With tires mounted upon them, they support the entire weight of the aircraft

during taxi, takeoff, and landing. The typical aircraft wheel is lightweight, strong, and made from aluminum alloy. Some magnesium alloy wheels also exist. Early aircraft wheels were of single piece construction, much the same as the modern automobile wheel. As aircraft tires were improved for the purpose they serve, they were made stiffer to better absorb the forces of landing without blowing out or separating from the rim. Stretching such a tire over a single piece wheel rim was not possible. A two-piece wheel was developed. Early two-piece aircraft wheels were essentially one-piece wheels with a removable rim to allow mounting access for the tire. These are still found on older aircraft. Later, wheels with two nearly symmetrical halves were developed. Nearly all modern aircraft wheels are of this two-piece construction.

[13] The typical modern two-piece aircraft wheel is cast or forged from aluminum or magnesium alloy. The halves are bolted together and contain a groove at the mating surface for an o-ring, which seals the rim since most modern aircraft utilize tubeless tires. The bead seat area of a wheel is where the tire actually contacts the wheel. It is the critical area that accepts the significant tensile loads from the tire during landing. To strengthen this area during manufacturing, the bead seat area is typically rolled to prestress it with a compressive stress load.

[14] **Aircraft brakes**: Very early aircraft has no brake system to slow and stop the aircraft while it is on the ground. Brake systems designed for aircraft became common after World War I as the speed and complexity of aircraft increased, and the use of smooth, paved runway surfaces proliferated. All modern aircraft are equipped with brakes. Their proper functioning is relied upon for safe operation of the aircraft on the ground. The brakes slow the aircraft and stop it in a reasonable amount of time. They hold the aircraft stationary during engine run-up and, in many cases, steer the aircraft during taxi. On most aircraft, each of the main wheels is equipped with a brake unit. The nose wheel or tail wheel does not have a brake.

[15] Modern aircraft typically uses disc brakes. The disc rotates with the turning wheel assembly while a stationary caliper resists the rotation by causing friction against the disc when the brakes are applied. The size, weight, and landing speed of the aircraft influence the design and complexity of the disc brake system. Single, dual, and multiple disc brakes are common types of brakes. The use of carbon discs is increasing in the modern aviation fleet.

[16] Large, heavy aircraft require the use of multiple-disc brakes. Multiple-disc brakes are heavy duty brakes designed for use with power brake control valves. The brake assembly consists of an extended bearing carrier similar to a torque tube type unit that bolts to the axle flange. The steel stators are keyed to the bearing carrier, and the rotors are keyed to the rotating wheel. Hydraulic pressure applied to the pistons causes the entire stack of stators and rotors to be compressed. This creates enormous friction and heat and slows the rotation of the wheel.

[17] Fig. 12-4 illustrates a brake assembly used on a B737 transport category aircraft. The carrier assembly is the basic unit of the brake. It is bolted to the shock strut axle flange with mount flange and serves as housing for actuating pistons. Regulated hydraulic pressure is applied through brake hose connection to the piston chambers in the carrier and forces the pistons to move outward, compressing the stators and the rotors. The resulting friction causes a braking action on the wheel and tire assembly. When the hydraulic pressure is relieved, the retracting springs in the automatic adjusters force the pistons to retract into the chambers and the hydraulic fluid is bled through the brake hose connection to return line. The automatic adjusters give correct clearances between the rotors and stators.

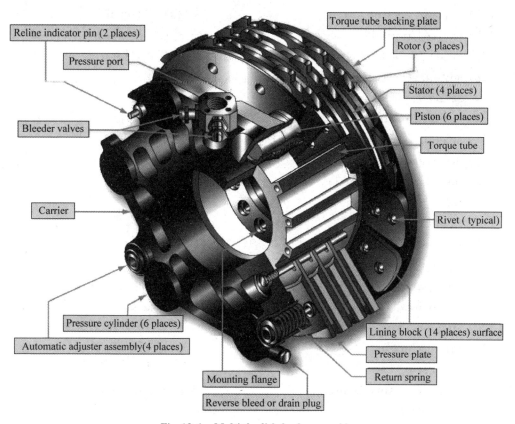

Fig. 12-4 Multiple-disk brake assembly

NEW WORDS AND EXPRESSIONS

(1) landing gear　　　　　　　　　　起落架
(2) wheel['wi:l]　　　　　　　　　　机轮
(3) airport['eəpɔ:t]　　　　　　　　机场
(4) runway['rʌnweɪ]　　　　　　　　跑道

(5) skid[skɪd] 尾撬,陆上飞机滑橇
(6) helicopter[ˈhelɪkɔptə] 直升机
(7) balloon gondola 热气球
(8) ski[skiː] 滑橇
(9) pontoon[pɔnˈtuːn] 浮筒
(10) shock absorbing 减震
(11) retraction mechanisms 收放机构
(12) cowling[ˈkaʊlɪŋ] 整流罩
(13) fairing[ˈfeərɪŋ] 整流罩
(14) conventional[kənˈvenʃn] 传统的
(15) tandem[ˈtændəm] 串列的,自行车式
(16) tricycle[ˈtraɪsɪkl] 前三点式
(17) arrangement[əˈreɪndʒmənt] 布局,配置,排列
(18) advantageous[ˌædvənˈteɪdʒəs] 优越的,优点
(19) sailplane[ˈseɪlpleɪn] 水上飞机
(20) outrigger[ˈaʊtrɪɡə(r)] 尾撑,承力支架
(21) steering[ˈstɪərɪŋ] 转弯,操纵,驾驶
(22) differential[ˌdɪfəˈrenʃl] 差动的
(23) mechanical[məˈkænɪkl] 机械的
(24) tiller[ˈtɪlə(r)] 手轮,舵柄
(25) flight deck 驾驶舱
(26) reinforced[riːɪnˈfɔːst] 加强,加固
(27) margin[ˈmɑːdʒɪn] 余量;裕度
(28) strut[strʌt] 支柱
(29) bogie[ˈbəʊɡɪ] 小车架,轮架
(30) alignment[əˈlaɪnmənt] 对齐
(31) priority[praɪˈɔrətɪ] 优先,优先权
(32) stow[stəʊ] 收起,贮藏
(33) emergency[ɪˈməːdʒənsɪ] 紧急的;紧急情况
(34) drum[drʌm] 鼓轮
(35) signal[ˈsɪɡnəl] 信号;发出信号
(36) spindle[ˈspɪndl] 轴
(37) compensator[ˈkɔmpenseɪtə] 补偿器
(38) tire[ˈtaɪə] 轮胎
(39) mount[maʊnt] 安装
(40) magnesium[mæɡˈniːzɪəm] 镁
(41) rim[rɪm] 边缘,轮缘
(42) symmetrical[sɪˈmetrɪkl] 对称地
(43) cast[kɑːst] 铸造

(44) forged[ˈfɔːdʒd]　　　　　　　　锻造
(45) groove[gruːv]　　　　　　　　沟,槽
(46) ring[rɪŋ]　　　　　　　　　　密封圈
(47) tubeless[ˈtjuːbləs]　　　　　　无内胎的
(48) bead[biːd]　　　　　　　　　胎圈
(49) tensile load　　　　　　　　　拉伸载荷
(50) rolled[rəʊld]　　　　　　　　滚压
(51) prestress[ˈpriːˈstres]　　　　 预应力
(52) paved[peɪvd]　　　　　　　　有铺装面的
(53) proliferated[prəʊˈlɪfəreɪtɪd]　 激增；猛增
(54) stationary[ˈsteɪʃənrɪ]　　　　 静止的
(55) caliper[ˈkælɪpə]　　　　　　　卡钳,卡尺
(56) bearing carrier　　　　　　　 轴承座
(57) torque[tɔːk]　　　　　　　　 扭矩,扭力
(58) axle[ˈæksl]　　　　　　　　　 轴
(59) flange[flændʒ]　　　　　　　 法兰盘,凸缘
(60) backplate[ˈbækpleɪt]　　　　 背板
(61) retainer[rɪˈteɪnə(r)]　　　　　保持器,止动器
(62) stack[stæk]　　　　　　　　　一叠,一摞,一堆

EXERCISE

1. Translate the followings into Chinese.

(1) **Ski plane**. An airplane whose wheeled landing gear has been replaced with skis so it can operate from snow or ice. Some types of skis require the wheels to be removed, and others can be installed without removing the wheels. Retractable skis have a slot in their bottom surface so they can be pulled up enough for the wheel to protrude below the ski, enabling the airplane to land on a hard surface on its wheels.

(2) **Brake**. A mechanism inside an aircraft wheel used to apply friction to the wheel to slow or stop its rotation.

(3) **Cowling**. The removable cover which encloses an aircraft engine.

(4) **Fairing**. A part of a structure or machine whose primary purpose is to produce a smooth surface or a smooth junction where two surfaces join.

(5) **Cylinder**. The component in a reciprocating engine which houses the piston and valves. The cylinders form the combustion chambers of the engine.

(6) **Compensator port**. A small hole between a hydraulic brake master cylinder and the reservoir. The first movement of the piston in the master cylinder covers

the compensating port and traps fluid between the master cylinder piston and the brake wheel cylinders to apply the brakes. When the brake is released, the piston uncovers the compensator port, and if any fluid has been lost from the brake, the reservoir will refill the master cylinder. A restricted compensator port may cause the brakes to drag or slow their release.

(7) **Bogie landing gear**. The landing gear of an aircraft that uses tandem wheels mounted along the center line of the aircraft fuselage. Some aircraft having bogie landing gear are supported while parked by outrigger wheels mounted far out on the wing.

(8) **Flange**. A ridge that protrudes from a device and is used for attaching something to the device, or for connecting two objects together. The propeller shaft of most modern aircraft engines is fitted with a flange on which the propeller mounts.

(9) **Metering pin**. A flow control device in a metering orifice, used to vary the amount of fluid that can flow through it. The area of the orifice with the pin fully in it determines the minimum amount of fluid that can flow, and the area of the jet with the pin all the way out determines the maximum amount of flow. The shape, or contour, of the metering pin determines the amount of fluid that can flow with the pin in any position other than full in or full out.

(10) **Tandem**. One object ahead of another. Tandem seating in an airplane has one seat ahead of the other. The wheels of a bicycle are tandem wheels.

2. **Answer the following questions in English.**
(1) What elements are included in the landing gear system?
(2) Why is the landing gear with tail wheel called conventional gear?
(3) What are the technical characteristics of the tail-type landing gear?
(4) What are the technical characteristics of the tricycle-type landing gear?
(5) Why are the main gears equipped with two or more wheels?
(6) What elements are included in the landing gear retraction system?
(7) How many control modes are there for aircraft maneuver on the ground?
(8) What elements are included in the aircraft brake assembly?

SUPPLEMENTARY READING

[1] **Shock struts**: True shock absorption occurs when the shock energy of landing impact is converted into heat energy, as in a shock strut landing gear. This is the most common method of landing shock dissipation in aviation. It is used on aircraft of all sizes. Shock struts are self-contained hydraulic units that support an aircraft while on the ground and protect the structure during landing. They must be inspected and serviced

regularly to ensure proper operation. There are many different designs of shock strut, but most operate in a similar manner.

[2] A typical pneumatic/hydraulic shock strut uses compressed nitrogen combined with hydraulic fluid to absorb and dissipate shock loads. It is sometimes referred to as an air/oil or oleo strut. A shock strut is constructed of two telescoping cylinders or tubes that are closed on the external ends. The upper cylinder is fixed to the aircraft and does not move. The lower cylinder is called the piston and is free to slide in and out of the upper cylinder. Two chambers are formed. The lower chamber is always filled with hydraulic fluid and the upper chamber is filled with compressed nitrogen. An orifice located between the two cylinders provides a passage for the fluid from the bottom chamber to enter the top cylinder chamber when the strut is compressed.

[3] Most shock struts employ a metering pin for controlling the rate of fluid flow from the lower chamber into the upper chamber. During the compression stroke, the rate of fluid flow is not constant. It is automatically controlled by the taper of the metering pin in the orifice. When a narrow portion of the pin is in the orifice, more fluid can pass to the upper chamber. As the diameter of the portion of the metering pin in the orifice increases, less fluid passes. Pressure build-up caused by strut compression and the hydraulic fluid being forced through the metered orifice causes heat. This heat is converted from impact energy. It is dissipated through the structure of the strut.

[4] On some types of shock strut, a metering tube is used. The operational concept is the same as that in shock struts with metering pin, except the holes in the metering tube control the flow of fluid from the bottom chamber to the top chamber during compression.

[5] Upon lift off or rebound from compression, the shock strut tends to extend rapidly. This could result in a sharp impact at the end of the stroke and damage to the strut. It is typical for shock struts to be equipped with a damping or snubbing device to prevent this. A recoil valve on the piston or a recoil tube restricts the flow of fluid during the extension stroke, which slows the motion and prevents damaging impact forces.

[6] Most shock struts are equipped with an axle as part of the lower cylinder to provide installation of the aircraft wheels. Shock struts without an integral axle have provisions on the end of the lower cylinder for installation of the axle assembly. Suitable connections are provided on all shock strut upper cylinders to attach the strut to the airframe.

[7] The compression stroke of the shock strut begins as the aircraft wheels touch the ground. As the center of mass of the aircraft moves downward, the strut compresses, and the lower cylinder or piston is forced upward into the upper cylinder. The metering pin is therefore moved up through the orifice. The taper of the pin controls the rate of fluid flow from the bottom cylinder to the top cylinder at all points during the

compression stroke. In this manner, the greatest amount of heat is dissipated through the walls of the strut. At the end of the downward stroke, the compressed nitrogen in the upper cylinder is further compressed which limits the compression stroke of the strut with minimal impact. During taxi operations, the air in the tires and the strut combine to smooth out bumps.

[8] **Brake system**: Large and high-performance aircraft are equipped with power brakes to slow, stop, and hold the aircraft. Power brake actuating systems use the aircraft hydraulic system as the source of power to apply the brakes. The pilot presses on the top of the rudder pedal for braking as with the other actuating systems. The volume and pressure of hydraulic fluid required cannot be produced by a master cylinder. Instead, a power brake control valve or brake metering valve receives the brake pedal input either directly or through linkages. The valve meters hydraulic fluid to the corresponding brake assembly in direct relation to the pressure applied to the pedal.

[9] Many power brake system designs are in use. Most are similar to the simplified system. Power brake systems are constructed to facilitate graduated brake pressure control, brake pedal feel, and the necessary redundancy required in case of hydraulic system failure. Large aircraft brake systems integrate anti-skid detection and correction devices. These are necessary because wheel skid is difficult to detect on the flight deck without sensors. However, a skid can be quickly controlled automatically through pressure control of the hydraulic fluid to the brakes. Hydraulic fuses are also commonly found in power brake systems. The hostile environment around the landing gear increases the potential for a line to break or sever, a fitting to fail, or other hydraulic system malfunctions to occur where hydraulic fluid is lost enroute to the brake assemblies. A fuse stops any excessive flow of fluid when detected by closing to retain the remaining fluid in the hydraulic system. Shuttle valves are used to direct flow from optional sources of fluid, such as in redundant systems or during the use of an emergency brake power source.

[10] Brake control valve/brake metering valve: the brake metering valves not only receive hydraulic pressure from two separate hydraulic systems, they also feed two separate brake assemblies. Each main wheel assembly has two wheels. The inboard wheel brake and the outboard wheel brake, located in their respective wheel rims, are independent from each other. In case of hydraulic system failure or brake failure, each is independently supplied to adequately slow and stop the aircraft without the other. More complicated aircraft may involve another hydraulic system for back-up or use a similar alternation of sources and brake assemblies to maintain braking in case of hydraulic system or brake failure.

[11] The parking brake system function is a combined operation. The brakes are applied with the rudder pedals and a ratcheting system holds them in place when the parking brake lever on the flight deck is pulled. At the same time, a shut-off valve is

closed in the common return line from the brakes to the hydraulic system. This traps the fluid in the brakes holding the rotors stationary. Depressing the pedals further releases the pedal ratchet and opens the return line valve.

[12] Anti-skid brake system: Large aircraft with power brakes require anti-skid systems. It is not possible to immediately ascertain in the flight deck when a wheel stops rotating and begins to skid, especially in aircraft with multiple-wheel main landing gear assemblies. A skid not corrected can quickly lead to a tire blowout, possible damage to the aircraft, and control of the aircraft may be lost.

[13] The anti-skid system not only detects wheel skid, it also detects when wheel skid is imminent. It automatically relieves pressure to the brake pistons of the wheel in question by momentarily connecting the pressurized brake fluid area to the hydraulic system return line. This allows the wheel to rotate and avoid a skid. Lower pressure is then maintained to the brake at a level that slows the wheel without causing it to skid.

[14] Maximum braking efficiency exists when the wheels are decelerating at a maximum rate but are not skidding. If a wheel decelerates too fast, it is an indication that the brakes are about to lock and cause a skid. To ensure that this does not happen, each wheel is monitored for a deceleration rate faster than a preset rate. When excessive deceleration is detected, hydraulic pressure is reduced to the brake on that wheel. To operate the anti-skid system, flight deck switches must be placed in the ON position. After the aircraft touches down, the pilot applies and holds full pressure to the rudder brake pedals. The anti-skid system then functions automatically until the speed of the aircraft has dropped to approximately 20 mph. The system returns to manual braking mode for slow taxi and ground maneuvering.

TOPIC 12：机务放行英语口语

表 T12-1 给出了飞机进出港时地面维护人员与机组人员交流中用到的常用口语示例。

表 T12-1　飞机进出港常用口语

编号	English	中文
1	Roger	收到且明白对方的意思
2	Chocks in/Position chocks	轮挡挡好
3	Remove chocks	撤除轮挡
4	Brakes off/Release parking brakes	松刹车/解除停留刹车
5	Brakes off/Parking brake released	刹车已松/停留刹车已解除
6	Brakes on/Set parking brake	刹车已刹好/设置停留刹车
7	Brakes on/Parking brake set	刹车已刹好/停留刹车已设置
8	Ground power connected	地面电源设备接好
9	Disconnect ground equipment	断开地面设备
10	Ground equipment disconnected	地面设备已断开
11	All doors checked closed	所有舱门都已关好
12	Pressure on	已经供气
13	Insert/Remove landing gear safety pin	插上/取下起落架安全销
14	Pin removed	插销已取下
15	Standby for pushback	稍等推出
16	Clear for pushback	可以牵引
17	Ready for pushback	准备牵引
18	Pushback complete	牵引完成
19	Standby	稍等
20	Standby for start	稍等启动
21	Start number one/or two	准备启动 1/2 号发动机
22	Clear number one/or two	可以启动 1/2 号发动机
23	Start complete	启动完成
24	Shut down engines	请关车
25	Runway two five/face two five	25 号跑道起飞
26	Runway zero three/face zero three	03 号跑道起飞
27	Hand signal on the left/or right	在左/右打手势
28	Acknowledge	明白(收到并听懂对方的话)
29	Affirmative	赞同(对方的建议)
30	Confirm	确认
31	Correct	正确
32	Go ahead	请继续(讲)
33	Ground check please	请绕机检查
34	Ground check completed	绕机检查完成
35	Hydraulic clearance please	请做好启动液压泵前检查

续表

编号	English	中文
36	Hydraulic clearance completed	启动液压泵检查完成
37	Read back (Say my message as you heard it)	重复你听到的信息
38	Say again	请重复你说的信息
39	Say again slowly	请重复,语速慢点

表 T12-2 给出了飞机出港机务人员与机组交流口语示例。以飞机正常出港为例:廊桥和勤务车辆撤离,绕飞机检查,确保所有舱门/盖板已经关好,等待推出,防撞灯亮起后,连接拖把,插上耳机与机组联系,准备推飞机。

表 T12-2 飞机出港机务人员与机组交流口语示例

Ground to Cockpit	Cockpit to Ground
Ground to cockpit 地面呼叫机组	Go ahead 请讲
Ground check complete, all doors closed and locked, Tractor and tow bar connected, we are ready for push back. 地面检查完毕,所有舱门已关上并锁好,牵引车和拖把已连接,我们已经做好推出准备	Roger. /Roger, standby. 收到。/收到,请稍候
Remove ground power unit? 可以断开地面电源吗?	Yes, ready for remove. 可以
Please release parking brake. 请解除停留刹车 Please confirm brakes released. 确认停留刹车已解除	Roger, parking brake released. 收到,停留刹车已经解除
Where nose toward? 机头朝哪个方向	Nose eastward. 机头朝东
Roger, we start pushback now. 收到,开始推出	

此外,在日常维修工作中,维修人员之间以及维修人员与地勤人员之间常用的口语示例如下:

(1) Could you please tell me the ETA of flight CA918? Nineteen thirty local time.
你能告诉我 CA981 航班预计到达时间吗? 当地时间 19:30。

(2) The flight is delayed due to the weather/mechanical fault.
航班因为天气/机械故障而延误了。

(3) Make sure the aircraft parking area is clean and clear of obstruction.
请确保飞机停机区域清洁,无障碍物。

(4) Please contact the cockpit with interphone.
请用内话与驾驶舱联系。

(5) The fuel control unit has been installed. Now we'll test the engines.
燃油控制器已装好,现在要试车检查。

(6) Request towing/pushback, it is not into position, a little more forward please.
请求拖/推飞机,没有进入位置,请向前移一点。

(7) The cross wind on the ground is so strong that the aircraft can not be towed.
地面侧风太大,所以不能拖飞机。

(8) There are ice on the ground, please slow down when turning.
地面结冰,转弯时请减速。

(9) Is there any trouble with the aircraft?
飞机有故障吗?
Everything is ok! /Yes, please look at flight log book/cabin log book.
一切正常。/是的,请看飞行记录本/客舱记录本。

(10) Please sign on the flight log book.
请在飞行记录本上签字。

(11) Please open nose/main landing gear compartment door and check the system line/wire/component for damage/loose/leakage.
请打开前/主起落架舱门,检查系统管路/导线/部件有无损坏/松动/渗漏。

(12) Shock absorber sliding tube extension is correct and the surface is clean.
减震器内筒伸出正常且表面洁净。

(13) Check the indication of pressure gauge on left main landing gear shock strut.
请检查左主起落架减震支柱压力指示。
The pressure indication is normal.
压力指示正常。

(14) Please check tyre for damage and wear.
请检查轮胎的损坏和磨损情况。

(15) Please check tyre pressure/condition.
请检查轮胎压力/状态。

(16) Please get me a compressed nitrogen cylinder to charge the tyre.
请拿一个压缩氮气瓶给轮胎充气。

(17) No damage and leakage for landing gear assembly structure attachment and up-assembly is found.
起落架组件结构连接以及上锁组件无损伤和渗漏现象。

(18) Check for the wheel rim damaged.
检查轮缘有无损伤。

(19) Check for the wheel sheared/missing tie bolts.
检查轮子拉紧螺栓有无剪断/缺失。

(20) Check brake unit for leakage/overheat.
检查刹车组件有无渗漏/过热现象。

(21) We are going to have the left wheel assembly replaced. Please get me two axle jacks.
我们要更换左轮组件,请帮助找两个轮轴千斤顶。

(22) Please check for brake assembly wear.
请检查刹车组件的磨损情况。

(23) Please wipe excess oil from engine nacelles and landing gears.
擦掉发动机短舱和起落架上多余的油。

(24) No damage and hydraulic fluid leakage from engine pylon and cowling doors was found.
发动机吊架和整流罩门未发现液压油渗漏和损伤。

(25) Please open the engine cowling and check for leakage/overheat/wires/electrical connectors.
请打开发动机整流罩检查有无渗漏/过热/导线/电插头(有无松开)。

(26) Check fan blades and turbine blades for damage.
检查风扇叶片和涡轮叶片有无损坏迹象。

(27) Check the engine tail pipe for metal particles and visible damage.
检查发动机尾喷管有无金属颗粒和可见损坏。

(28) Engine IDG oil needs Replenishment.
发动机 IDG 滑油需要补加。

(29) Main cargo compartment door can't be opened electrically, in order to avoid aircraft delay, we want to open it manually.
主货舱门无法电动打开,为了不延误飞机,我们想人工打开。

(30) Check cargo compartment lining for condition and cleaning.
检查货舱壁是否完好且干净。

(31) Please close cargo compartment doors.
请将货舱门关上。

(32) Drive the galley service truck slowly and mind the aircraft.
驾驶厨房服务车要慢点,注意不要碰坏飞机。

(33) Please check crew and passenger oxygen system pressure.
请检查机组和旅客氧气系统压力。

(34) Please drain off all potable water or it will be frozen.
请将饮用水系统的水全部放掉,否则会结冰。

(35) Please refill water for potable water system.
请给饮用水系统加水。

(36) Please arrange a water servicing unit/a toilet servicing unit/an air conditioning unit/a deicing unit/a ladder for us.
请为我们安排一辆加水车/污水车/空调车/除冰车/工作梯。

(37) Please connect the ground power unit/the air start unit.
请接上地面电源/气源车。

(38) APU is fully inoperative, please call a ground power unit and two air start units immediately.
APU 完全失效,请马上来一辆地面电源车和两辆气源车。

(39) Could you please provide us a sealing ring (an adjustable wrench, a round file, a triangle file, a wrench, a socket wrench, universal wrench, a cutting pliers, a rubber hammer, some split cotters, some safety wire, chamois leather, a piece of cleaning cloth, a bucket, a washer, screwdriver, clamps, pliers,

flashlight)?

请给我们提供一个密封圈(或活动扳手、圆锉、三角锉、扳手、套筒扳手、万用扳手、克丝钳、橡胶锤、一些开口销、一些保险丝、鹿皮布、一块抹布、油桶、垫片、螺丝刀、夹子、钳子、手电筒)。

(40) The aircraft needs to be refilled with hydraulic fluid/oil.

飞机需要加液压油/滑油。

(41) Please replenish two quarters of MOBIL JET 2$^{\#}$ oil for engine NO.1 and NO.2, three quarters for engine NO.3 and NO.4.

请给1、2号发动机加2夸脱美孚2$^{\#}$滑油,3、4号发动机加3夸脱滑油。

(42) It is not allowed to use this brand of oil in our aircraft.

这种牌号的滑油我们飞机不能用。

(43) The aircraft needs refueling/defuelling. Please call a refueling/defuelling-cart.

飞机需要加/抽燃油,请叫一辆加/抽油车。

(44) I want to check the fuel samples.

我要检查一下油样。

(45) Please put fire extinguisher in position before refueling.

加油前请将灭火瓶放置在指定位置。

(46) Please drain some deposit and water from No. 2 tank.

请给2号油箱放沉淀物和水。

(47) Please guide us to parking place.

请引导我们到停机位。

(48) Please connect/disconnect the tow-bar.

请连接/断开拖把。

UNIT 13

LIGHTING SYSTEM

[1] Aircraft lighting systems provide illumination for both exterior and interior use. Lights on the exterior provide illumination for such operations as landing at night, inspection of icing conditions, and safety from midair collision. Interior lighting provides illumination for instruments, cockpits, cabins, and other sections occupied by crewmembers and passengers. Certain special lights, such as indicator and warning lights, indicate the operational status of equipment.

[2] Position, anti-collision, landing, and taxi lights are common examples of aircraft exterior lights. Some lights, such as position lights and anti-collision lights, are required for night operations. Other types of exterior lights, such as wing inspection lights, are of great benefit for specialized flying operations.

[3] Aircraft operating at night must be equipped with position lights that meet the minimum requirements specified by the Civil Aviation Regulations. A set of position lights consist of one red, one green, and one white light. Position lights are sometimes referred to as "navigation" lights. On many aircraft each light unit contains a single lamp mounted on the surface of the aircraft ((a) of Fig. 13-1). Other types of position light units contain two lamps ((b) of Fig. 13-1), and are often streamlined into the surface of the aircraft structure. The green light unit is always mounted at the extreme tip of the right wing. The red unit is mounted in a similar position on the left wing. The white unit is usually located on the vertical stabilizer in a position where it is clearly visible through a wide angle from the rear of the aircraft.

[4] The wingtip lamps and the tail lamps are controlled by a double-pole, single-throw switch in the pilot's compartment. On "dim", the switch connects a resistor in series with the lamps since the resist or decreases current flow, the light intensity is reduced. On "bright", the resistor is shorted out of the circuit, and the lamps glow at full brilliance. On some types of installations a switch in the pilot's compartment provides for steady or flashing operation of the position lights. For flashing operation, a flasher mechanism is usually installed in the position light circuit. It consists essentially of a motor-driven camshaft on which two cams are mounted and a switching mechanism made up of two breaker arms and two contact screws. One breaker arm supplies DC

(a) Tail position light unit   (b) Wingtip position light unit

Fig. 13-1   Position light

current to the wingtip light circuit through one contact screw, and the other breaker arm supplies the tail light circuit through the other contact screw. When the motor rotates, it turns the camshaft through a set of reduction gears and causes the cams to operate the breaker which opens and closes the wing and taillight circuits alternately.

[5] There are, of course, many variations in the position light circuits used on different aircraft. All circuits are protected by fuses or circuit breakers, and many circuits include flashing and dimming equipment. Still others are wired to energize a special warning light dimming relay, which causes all the cockpit warning lights to dim perceptibly when the position lights are illuminated.

[6] Small aircrafts are usually equipped with a simplified control switch and circuitry. In some cases, one control knob or switch is used to turn on several sets of lights; for example, one type utilizes a control knob, the first movement of which turns on the position lights and the instrument panel lights. Further rotation of the control knob increases the intensity of only the panel lights. A flasher unit is seldom included in the position light circuitry of very light aircraft, but is used in small twin-engine aircraft.

[7] An anti-collision light system may consist of one or more lights. They are rotating beam lights which are usually installed on top of the fuselage or tail in such a location that the light will not affect the vision of the crew member or detract from the conspicuousness of the position lights. In some cases, one of the lights is mounted on the underside of the fuselage.

[8] The simplest means of installing an anti-collision light is to secure it to a reinforced fuselage skin panel, as shown in Fig. 13-2.

[9] An anti-collision light is often installed on top of the vertical stabilizer if the cross section of the stabilizer is large enough to accommodate the installation, and if aircraft flutter and vibration characteristics are not adversely affected. Such installations should be located near a spar, and formers should be added as required to stiffen the structure near the light. Fig. 13-3 shows a typical anti-collision light installation in a vertical stabilizer.

Fig. 13-2 Typical anti-collision light installation in an unpressurized skin

Fig. 13-3 Typical anti-collision light installation in a vertical stabilizer

[10] An anti-collision light unit usually consists of one or two rotating lights operated by an electric motor. The light may be fixed, but mounted under rotating mirrors inside a protruding red glass housing (Fig. 13-4). The mirrors rotate in an arc, and the resulting flash rate is between 40 and 100 cycles per minute. The anti-collision light is a safety light to warn other aircraft, especially in congested areas.

[11] Landing lights are installed in aircraft to illuminate runways during night landings. These lights are very powerful and are directed by a parabolic reflector at an angle providing a maximum range of illumination. Landing lights are usually located midway in the leading edge of each wing or streamlined into the aircraft surface. Each light may be controlled by a relay, or it may be connected directly into the electric circuit.

[12] Since icing of the lamp lenses reduces the illumination quality of a lamp, some installations use retractable landing lamps (Fig. 13-5). When the lamps are not in use, a motor retracts them into receptacles in the wing where the lenses are not exposed to the weather.

Fig. 13-4 Anti-collision light

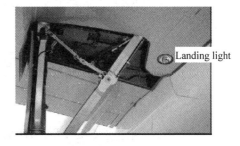

Fig. 13-5 Retractable landing light

[13] As shown in Fig. 13-6, one type of retractable landing light motor has a split-field winding. Two of the field winding terminals connect to the two outer terminals of

the motor control switch through the points of contacts "C" and "D", while the center terminal connects to one of two motor brushes. The brushes connect the motor and magnetic brake solenoid into the electric circuit. The points of contact "C" are held open by the geared quadrant of the landing lamp mechanism. The points of contact "D" are held closed by the tension of the spring to the right of the contacts. This is a typical arrangement of a landing lamp circuit when the landing lamp is retracted and the control switch is in the "OFF" position. No current flows in the circuit, and neither the motor nor the lamp can be energized. When the control switch is placed in the upper, or "extend" position, current from the battery flows through the closed contacts of the switch, the closed contacts of contact "D", the center terminal of the field winding, and the motor itself. Current through the motor circuit energizes the brake solenoid, which withdraws the brake shoe from against the motor shaft, allowing the motor to turn and lower the lamp mechanism. After the lamp mechanism moves about 10°, contact "A" touches and rides along the copper bar B. In the meantime, relay E is energized, and its contacts close. This permits current to flow through the copper bar B, contact "A", and

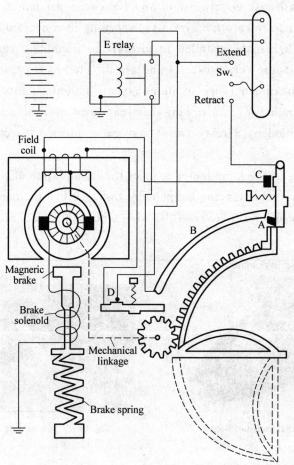

Fig. 13-6 Landing light mechanism and circuit

the lamp. When the lamp mechanism is completely lowered, the projection at the top of the gear quadrant pushes the "D" contacts apart, opens the circuit to the motor, and causes the de-energized brake solenoid to release the brake. The brake is pushed against the motor shaft by the spring, stopping the motor and completing the lowering operation.

[14] To retract the landing lamp, the control switch is placed in the "retract" position. The motor and brake circuits are completed through the points of contact "C", since these contacts are closed when the gear quadrant is lowered. This action completes the circuit, the brake releases, the motor turns (this time in the opposite direction) and the landing light mechanism is retracted. Since switching to "retract" breaks the circuit to relay E, the relay contacts open, disconnecting the copper bar and causing the landing lamp to go out. When the mechanism is completely retracted, contact points "C" open, and the circuit to the motor is again broken, the brake applied, and the motor stopped.

[15] Retractable landing lights that can be extended to any position of their extension are employed on some aircraft. Landing lights used on the high-speed aircraft are usually equipped with an airspeed pressure switch which prevents extension of landing lights at excessive airspeeds. Such switches also cause retraction of landing lights if the aircraft exceeds a predetermined speed.

[16] Many large aircraft are equipped with four landing lights, two of which are fixed and two retractable. Fixed lights are usually located in either the wing root areas or just outboard of the fuselage in the leading edge of each wing. The two retractable lights are usually located in the lower outboard surface of each wing, and are normally controlled by separate switches.

[17] On some aircraft, the fixed landing light is mounted in an area with a taxi light, as shown in Fig. 13-7. Taxi lights are designed to provide illumination on the ground while taxiing or towing the aircraft to or from a runway, taxi strip, or in the hangar area.

[18] Taxi lights are not designed to provide the degree of illumination necessary for landing lights; 150-250-watt taxi lights are typical on many medium and heavy aircraft. On aircraft with tricycle landing gear, either single or dual taxi lights are often mounted on the non-steerable part of the nose landing gear. They are positioned at an oblique angle to the center line of the aircraft to provide illumination directly in front of the aircraft and also some illumination to the right and left of the aircraft's path. On some aircraft the dual taxi lights are supplemented by wingtip clearance lights controlled by the same circuitry.

[19] Taxi lights are also mounted in the recessed areas of the wing leading edge, often in the same area with a fixed landing light.

[20] Many small aircraft are not equipped with any type of taxi light, but rely on the intermittent use of a landing light to illuminate taxiing operations. Still other

aircraft utilize a dimming resistor in the landing light circuit to provide reduced illumination for taxiing. A typical circuit for dual taxi lights is shown in Fig. 13-8.

Fig. 13-7　Fixed landing light and taxi light

Fig. 13-8　Taxi light mounted on the non-steerable portion of the nose landing gear

[21] Some large aircraft are equipped with alternate taxi lights located on the lower surface of the aircraft, aft of the nose radome. These lights, operated by a separate switch from the main taxi lights, illuminate the area immediately in front of and below the aircraft nose.

[22] Some aircraft are equipped with wing inspection lights to illuminate the leading edge of the wings to permit observation of icing and general condition of these areas in flight. On some aircraft, the wing inspection light system (also called wing ice lights) consists of a 100-watt light mounted flush on the outboard side of each nacelle forward of the wing. These lights permit visual detection of ice formation on wing leading edges while flying at night. They are also often used as floodlights during ground servicing. They are usually controlled through a relay by an "ON-OFF" toggle switch in the cockpit. Some wing inspection light systems may include or be supplemented by additional lights, sometimes called nacelle lights, that illuminate adjacent areas such as cowl flaps or the landing gear. These are normally the same type of lights and can be controlled by the same circuits.

NEW WORDS AND EXPRESSIONS

(1) transient['trænzɪənt]　　　　　　　　短暂的；瞬变现象
(2) voltage['vəʊltɪdʒ]　　　　　　　　　电压
(3) conductive[kən'dʌktɪv]　　　　　　 导电的
(4) illumination[ɪˌluːmɪ'neɪʃn]　　　　　照明；亮度
(5) exterior[ɪk'stɪərɪə(r)]　　　　　　　外部的
(6) interior[ɪn'tɪərɪə(r)]　　　　　　　　内部的
(7) crewmember[kruː'membə]　　　　　机组成员

(8) indicator['ɪndɪkeɪtə(r)] 指示器
(9) warning['wɔːnɪŋ] 警告
(10) anti-collision 防撞的
(11) inspection[ɪn'spekʃn] 检查
(12) lamp[læmp] 灯
(13) extreme[ɪk'striːm] 极度的；极端
(14) position[pə'zɪʃn] 位置
(15) visible['vɪzəbl] 可见的
(16) wingtip['wɪŋtɪp] 翼尖
(17) dim[dɪm] 变暗，暗亮
(18) connect[kə'nekt] 连接
(19) resistor[rɪ'zɪstə(r)] 电阻器
(20) in series with 串联
(21) current flow 电流
(22) intensity[ɪn'tensəti] 强度
(23) short out 短路
(24) glow[gləʊ] 发光的
(25) brilliance['brɪliəns] 明亮；亮度
(26) camshaft['kæmʃɑːft] 凸轮轴
(27) mechanism['mekənɪzəm] 机械装置
(28) screw[skruː] 螺钉
(29) alternately[ɔːl'tɜːnətlɪ] 交替地
(30) fuse[fjuːz] 保险丝
(31) breaker['breɪkə(r)] 跳开关，断路器
(32) relay['riːleɪ] 继电器
(33) illuminate[ɪ'luːmɪneɪt] 照明；照亮
(34) vision['vɪʒn] 视力；视觉
(35) conspicuousness[kən'spɪkjʊəsnəs] 显著，卓越
(36) accommodate[ə'kɒmədeɪt] 容纳；调节
(37) flutter['flʌtə(r)] 颤振
(38) vibration[vaɪ'breɪʃn] 振动
(39) stiffen['stɪfn] 加强
(40) congested[kən'dʒestɪd] 拥挤的
(41) lens[lenz] 透镜；镜头

EXERCISE

1. Translate the followings into Chinese.

(1) **Position light**. A position light is known as navigation light that a colored source

of illumination on a waterborne vessel, aircraft and some spacecraft, used to signal a craft's position, heading, and status. Commonly, their placement is mandated by international conventions or civil authorities.

(2) **Crewmember**. A person aboard an aircraft for the purpose of operating the aircraft in flight. **14 CFR Part 1**: "A person assigned to perform duty in an aircraft during flight time."

(3) **Flasher**. On aircraft primarily, flashers flash a high-intensity burst of white light, to help other pilots recognize the aircraft's position in low-visibility conditions.

(4) **Breaker**. A device that automatically interrupts the flow of electricity (current) during fault condition to protect wiring from excess current. Circuit breakers are also used as a means to manually deactivate the system.

(5) **Toggle switch**. A toggle switch is a class of electrical switches that are manually actuated by a mechanical lever, handle, or rocking mechanism.

(6) **Rheostat**. An electrical device for changing the resistance in an electrical circuit.

(7) **Flutter**. Flutter is a self-feeding and potentially destructive vibration where aerodynamic forces on an object couple with a structure's natural mode of vibration to produce rapid periodic motion. Flutter can occur in any object within a strong fluid flow, under the conditions that a positive feedback occurs between the structure's natural vibration and the aerodynamic forces. That is, the vibrational movement of the object increases an aerodynamic load, which in turn drives the object to move further.

(8) **Hangar**. A hangar is a closed structure to hold aircraft or spacecraft in protective storage. Most hangars are built of metal, but other materials such as wood and concrete are also used. Hangars are used for: protection from weather, protection from direct sunlight, maintenance, repair, manufacture, assembly and storage of aircraft on airfields.

(9) **Floodlight**. A floodlight is an artificial light providing even illumination across a wide area.

(10) **In series with**. A series of components or networks, the output of each of which serves as the input for the next.

2. Answer the following questions in English.
(1) Please describe the main functions of the aircraft lighting system.
(2) Please describe the two parts of the aircraft lighting system.
(3) Please describe the main constitutes of the Interior aircraft lighting system.
(4) Please describe the main constitutes of the exterior aircraft lighting system.
(5) Please describe the two kinds of the aircraft landing lights.

SUPPLEMENTARY READING

[1] The purpose of the lighting system is to provide the necessary illumination for passenger comfort, optimum flight crew work performance, service and cargo handling, and provide for lighting under emergency conditions. The major systems are: flight compartment lights, passenger compartment lights, exterior lights and emergency lights. Here, we will introduce the flight deck lighting only.

[2] Variable intensity controls for the integral instrument and main instrument panel lights are identified as "PANEL" and located on the lower portion of the captain's and first officer's panels. The captain's variable intensity "PANEL" switch controls these lights on the captain's main panel, the center panel and the glareshield. The first officer's "PANEL" switch controls these lights on the first officer's panel.

[3] White fluorescent floodlights are positioned under the glareshield to direct background lighting onto the main panels. One variable-intensity switch, which controls these lights, is on the lower portion of the captain's main panel and identified as "BACKGROUND", movement of this switch in a clockwise direction increases the brightness of the fluorescent lights. White incandescent floodlights are positioned above the autopilot flight director system (AFDS) panel, one variable-intensity switch, which controls these lights, is on the lower portion of the captain's main panel and identified as "AFDS FLOOD". Clockwise movement of this switch increases brightness of the lights.

[4] The flight compartment system indicator lights may be dimmed or tested by the use of a "LIGHT" switch located on the center instrument panel. Placing the switch in the test position illuminates the indicator lights, and placing the switch in the dim position dims them.

[5] The control stand instrument panels are integrally lighted similar to those units on the main instrument panels. In addition, there is a white flood light on the forward overhead panel which directs light downward onto the thrust lever quadrant on the control stand. Variable-intensity controls for these lights are on the control stand and identified as "PANEL" and "FLOOD".

[6] A variable intensity control for the integral instrument panel lights and for instrument lights on the overhead panels is located on the forward overhead panels and identified as "PANEL".

[7] In addition, there are two map lights mounted above each pilot's seat, two circuit breaker panel floodlights on the sidewall by each pilot's seat, two white dome lights on the circuit breaker panels behind the crew provide general flight compartment area illumination.

TOPIC 13：飞机主要维修工具

飞机典型的通用维修工具如下：

安全带 life belt(safety belt)	焊锡 soldering tin	面罩 shield face
扳手 wrench	护目镜 goggle（blinkers）	抹布 dishcloth
半圆锉 half round file	护膝 kneepad	内六角扳手 hex wrench
保险钳 wire twister	划刀 utility knife	镊子 nipper（plier）
保险丝钳 safety wire twister	滑油加油车 oil cart	平板棘轮扳手 slab ratchet
表 gauge	活动扳手 adjustable wrench	平锉 flat file
剥线钳 wire stripper	棘轮扳手 ratchet wrench	平凿 flat chisel
插头钳 connector plug	加力杆 breaker bar	气枪 pneumatic runner
超声波 ultrasonic	尖嘴钳 needle nose plier	气钻 pneumatic drill
冲击工具 impact driver	剪刀 snip（scissor）	千斤顶 jack
充电器 charger	剪线钳 diagonal cutter	钳子 plier
磁性螺刀 magnetic screwdriver	胶带 adhesive tape	撬棒 tommy bar
锉 file	橡胶榔头 rubber hammer	全罩呼吸器 full face respirator
打磨片 grinding discs	万用刀具 multi-purpose cutter	软杆磁棒 flexible magnetic pick up tool
大力钳 locking plier	锯 saw	塞尺 feeler（thickness）gauge
电池 battery	卷尺 tape ruler	三脚架 tripod（spider）
电烙铁 electronic iron	卡尺 caliper	三用表 multi meter
电气工具 electronic tool	卡环钳 convertible snap ring plier	砂轮 grinding wheel
电枪 electronic runner	卡拉 ratchet	砂纸 sand paper
电源插座 receptacle	卡拉螺刀 ratcheting magnetic screwdriver	绳索 rope
电钻 electronic drill	开口/梅花扳手 combination wrench	十字头螺刀 pillips screwdriver
短螺刀 stubby screwdriver	开口扳手 open end wrench	手电 flashing light
对讲机 interphone	块扳手 crowfoot wrench	手套 glove
耳机 earphone	快速螺刀 fast screwdriver	锁定夹 locking clamp
耳塞 ear plug style protector	矿灯 safety lamb（miner's lamb）	套筒 socket
耳罩 headphone style ear protector	力矩扳手 torque wrench	套筒扳手 socket wrench
发光背心 flashing vest	链条扳手 chain wrench	套筒接杆 socket extension
反光镜 inspection mirror	量杯 measuring cup	铁榔头 iron hammer
防静电手腕 static wrist	六角套筒 six point socket	铜榔头 copper hammer
放大镜 magnifier	轮挡 wheel chock	拖把 tow bar
封圈拔具 seal puller	轮胎充气管 dial-type tire inflator	弯钩 hook
钢板尺 steel ruler	螺刀 screwdriver	弯头梅花扳手 box wrench
钢丝刷 wire brush	螺母螺刀 nut screwdriver	万向套筒 universal socket
工具架 tools rack	螺旋卡尺 outside micrometer	万向转接头 universal joint
工具箱 tools kit	毛刷 tapered head brush	香槟锤 dead blow hammer
工作梯 work ladder	铆钉枪 rivet gun	橡胶锤 mallet
刮刀 scraper knife	梅花扳手 spline end wrench	销钉插拔工具 pin removal tool
管钳 pipe wrench	梅花套筒 twelve point socket	指挥棒 flashing emergency light
压线钳 wire crimper	游标卡尺 dial caliper	中心冲 center punch
鸭嘴钳 duck bill plier	鱼口钳 combination slip-joint plier	注射器 injector（squirt）
扬声器 speaker	圆锉 round file	注油壶 oiler
摇把 speeder	圆头锤 ball peen hammer	注油枪 grease gun
液压油加油车 hydraulic fluid cart	杂项 diversiform	注油嘴 grease-fitting coupler
一字头螺刀 flat tip screwdriver	錾子 punch	转接头 adaptor
仪表螺刀 electronic miniature screwdriver	凿子 chisel	
鹰嘴钳 adjustable joint plier	兆欧表 meg ohm meter	

UNIT 14

OXYGEN SYSTEM

[1] The purpose of the oxygen system is to provide oxygen for the flight crew and passengers when required. As altitude increases, the air thins out and air pressure decreases. As a result, the amount of oxygen available to support life functions decreases. Aircraft oxygen systems are provided to supply the required amount of oxygen to keep a sufficient concentration of oxygen in the lungs to permit normal activity up to indicated altitudes of about 40,000 feet.

[2] There are three forms of oxygen on aircraft: gaseous oxygen, liquid oxygen, and chemical or solid oxygen. Most of the aircraft in the general aviation fleet use gaseous oxygen, usually stored in steel cylinders under a pressure of between 1,800 and 2,400 psi. Almost all military aircraft now carry their oxygen in its liquid. Liquid oxygen is a pale blue, transparent liquid that will remain in its liquid state as long as it is stored at a temperature of below $-181°F$. Chemical oxygen generators can be used to fulfill the new requirements. The chemical oxygen generator differs from the compressed oxygen cylinder and the liquid oxygen converter in that the oxygen is actually produced at the time of delivery. Solid state describes the chemical source, sodium chlorate, formula $NaClO_3$. When heated to $478°F$, sodium chlorate releases up to 45 percent of its weight as gaseous oxygen.

[3] There are three independent oxygen systems on modern airplane: The flight crew oxygen system is used to flight crew only. The passenger oxygen system is a chemically-generated oxygen system for use by passengers and cabin attendants. The portable oxygen system supply oxygen for emergency, first aid or sustaining oxygen for the passengers and/or crew.

[4] **The flight crew oxygen system**: The system is a diluter-demand type system which supplies oxygen to the flight crew, if there is a sudden decrease in cabin pressurization. It also supplies oxygen if there is smoke or dangerous gases in the cockpit. Each crew station has a quick-donning mask with a demand regulator installed. The oxygen is supplied from a high-pressure oxygen cylinder to the masks (through a pressure regulator/transmitter assembly and a distribution circuit). The system consists of an oxygen cylinder, valves, indicating components, distribution tubing and crew oxygen

mask/regulator (Fig. 14-1). The crew oxygen system components are located in the forward cargo compartment and flight compartment. Each crew member station is equipped with oxygen mask/regulator. The flight compartment oxygen system can provide air mixed with oxygen for the flight crew.

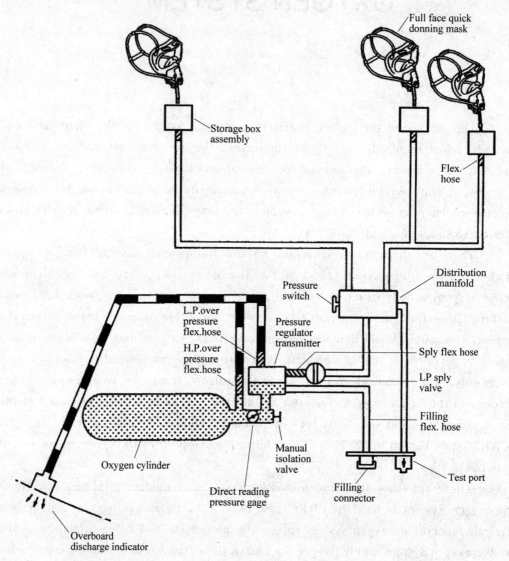

Fig. 14-1 The flight crew oxygen system

[5] **Passenger oxygen system**: If there is a sudden loss of cabin pressurization, emergency oxygen container supplies emergency oxygen to the passengers and to cabin attendants. Each chemical oxygen unit is made up of a solid-state chemical oxygen generator and two or more continuous-flow oxygen masks. Flexible supply hoses connect the masks to the chemical oxygen generators. The chemical oxygen units are installed: ①above the passenger seats; ②in the lavatories; ③at the cabin attendants stations; ④at the cabin attendants working areas.

[6] When the cabin pressure decreases to a pressure equivalent of 4,260 m (14,000 ft), the altitude pressure switch will close. This action operates the emergency oxygen system automatically. If the altitude pressure switch fails, you can operate the system with the "MASK MAN ON" pushbutton switch. This guarded electrical switch is installed in the cockpit overhead panel. The mask can also be deployed manually at any passenger service unit (PSU). The emergency oxygen flows when you pull the oxygen mask towards your face. This starts the chemical oxygen generator which gives approximately 15 minutes supply of oxygen.

[7] **Portable oxygen equipment**: Typical portable oxygen equipment consists of a lightweight steel alloy oxygen cylinder fitted with a combined flow control/reducing valve and a pressure gauge. A breathing mask, with connecting flexible tube and a carrying bag with the necessary straps for attachment to the wearer, completes the set. The charged cylinder pressure is usually 1,800 psi; however, the cylinder capacities vary. A popular size for portable equipment is the 120 liter capacity cylinder. Depending on the type of equipment used, it is normally possible to select at least two rates of flow, normal or high. With some equipment three flow rate selections are possible, i.e., normal, high, and emergency, which would correspond to 2, 4, and 10 liters per minute. With these flow rates a 120 liter cylinder would last for 60, 30, and 12 minutes, respectively.

[8] **Oxygen cylinder**: The oxygen is contained in either high pressure or low pressure oxygen cylinders. The high pressure cylinders are manufactured from heat treated alloy, or are wire wrapped on the outside surface, to provide resistance to shattering. All high pressure cylinders are identified by their green color and have the words "AVIATORS' BREATHING OXYGEN" stenciled lengthwise in white, 1 inch letters.

[9] High pressure cylinders are manufactured in a variety of capacities and shapes. These cylinders can carry a maximum charge of 2,000 psi, but are normally filled to a pressure of 1,800 to 1,850 psi.

[10] There are two basic types of low pressure oxygen cylinders. One is made of stainless steel; the other, of heat treated, low alloy steel. Stainless steel cylinders are made nonshatterable by the addition of narrow stainless steel bands that are seam welded to the body of the cylinder. Low alloy steel cylinders do not have the reinforcing bands but are subjected to a heat treatment process to make them nonshatterable. They have a smooth body with the word "NONSHATTERABLE" stenciled on them. Both types of low pressure cylinders come in different sizes and are painted light yellow. This color indicates that they are used for low pressure oxygen only. The cylinders may carry a maximum charge of 450 psi, but are normally filled to a pressure of from 400 to 425 psi. When the pressure drops to 50 psi, the cylinders are considered empty.

[11] **Oxygen generator**: Fig. 14-2 illustrates a schematic representation of a basic oxygen generator. The center axial position is occupied by a core of sodium chlorate,

iron, and some other ingredients mixed together and either pressed or cast into a cylindrical shape. This item has been popularly referred to as an oxygen candle, because when it is ignited at one end, it burns progressively in much the same manner as a candle or flare. Surrounding the core is porous packing. It supports the core and filters salt particles from the gas as it flows toward the outlet. A chemical filter and particulate filter at the outlet end of the container provide final clean-up of the gas so that the oxygen delivered is medically pure breathing oxygen. An initiation device is an integral part of the package. This may be either a mechanical percussion device or an electric squib. The choice depends on the application. The entire assembly is housed in a thin shelled vessel. Often included is a layer of thermal insulation on the inside shell, a check valve on the outlet, and a relief valve to protect against an inadvertent overpressure condition.

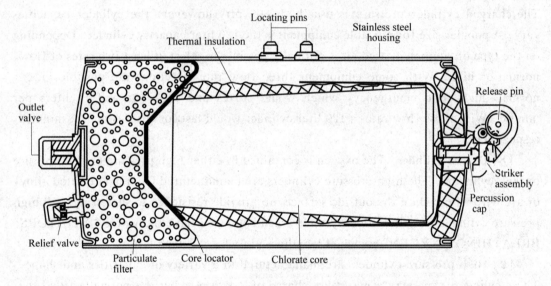

Fig. 14-2　Basic oxygen generator

[12] In operation, the burning is initiated at one end of the core by activating the squib or percussion device. Oxygen evolution rate is proportional to the cross-sectional area of the core and the burn rate. The burn rate is determined by the concentration of fuel in the chlorate. In certain cases, one end of the core is larger than the other. The purpose of this is to program a high oxygen evolution rate during the initial few minutes of burning such as is required for an emergency descent supply. Burning continues until the core is expended. There are no ON/OFF valves and no mechanical controllers. Refill is accomplished by simply replacing, in total, the entire device. A limitation is that once the generator is initiated, flow is delivered at a predetermined rate, thus demand use is not very efficient. In order to keep the process from consuming a great quantity of the oxygen, the quantity of iron is kept to a minimum. There is a tendency toward liberation of small amounts of chlorine. Barium peroxide, or barium dioxide,

may be added by the manufacturer to provide an alkaline medium for removing the trace amounts of chlorine that may be present.

[13] **Diluter demand regulators**: The diluter demand regulator gets its name from the fact that it delivers oxygen to the user's lungs in response to the suction of his own breath. To prolong the duration of the oxygen supply, the oxygen is automatically diluted in the regulator with suitable amounts of atmospheric air (Fig. 14-3). This dilution takes place at all altitudes below 34,000 feet.

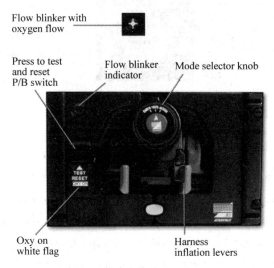

Fig. 14-3 Oxygen regulator

[14] The essential feature of a diluter demand regulator is a diaphragm operated valve called the demand valve, which opens by slight suction on the diaphragm during inhalation and which closes during exhalation. A reducing valve upstream from the demand valve provides a controlled working pressure. Downstream from the demand valve is the diluter control closing mechanism. This consists of an aneroid assembly (a sealed, evacuated bellows) which controls the air inlet valve. When the diluter lever is set in the position marked "normal oxygen", atmospheric air at ground level is supplied with very little oxygen added. As altitude increases, the air inlet is gradually closed by the bellows to give a higher concentration of oxygen until at about 34,000 feet the air inlet is completely closed and 100 percent oxygen is supplied. As altitude decreases, this process is reversed.

[15] Another type of diluter demand regulator is the narrow panel type. This type regulator face displays a float type flow indicator which signals oxygen flow through the regulator to the mask. The regulator face also displays three manual control levers. A supply lever opens or closes the oxygen supply valve. An emergency lever is used to obtain oxygen under pressure. An oxygen select lever is used for selecting an air/oxygen mixture or oxygen only. Fig. 14-4 illustrates how the narrow panel oxygen regulator operates. With the supply lever in the "ON" position, the oxygen select lever in the

"normal" position, and the emergency lever in the "OFF" position, oxygen enters the regulator inlet.

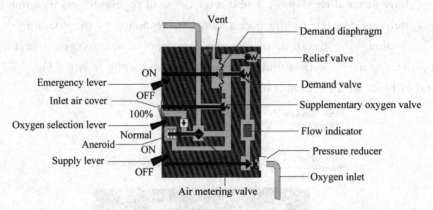

Fig. 14-4 Schematic of operation

[16] When there is sufficient differential pressure across the demand diaphragm, the demand valve opens to supply oxygen to the mask. This pressure differential exists during the user's inhalation cycle. After passing through the demand valve, the oxygen is mixed with air that enters through the air inlet port. The mixture ratio is determined by an aneroid controlled air metering valve. A high oxygen ratio is provided at high altitudes and a high air ratio at lower altitudes. The air inlet valve is set to permit the airflow to begin at the same time as the oxygen flow.

[17] The addition of air may be cut by turning the oxygen select lever to "100 percent". When this lever is in "normal", air enters through the air inlet port, and the required amount is added to the oxygen to form the correct air/oxygen mixture. Positive pressure at the regulator outlet may be obtained by turning the emergency lever to "ON". This mechanically loads the demand diaphragm to provide positive outlet pressure.

[18] **Oxygen mask**: There are numerous types of oxygen masks in use which vary widely in design detail. It is important that the masks used be compatible with the particular oxygen system involved. In general crew masks are fitted to the user's face with a minimum of leakage. Crew masks usually contain a microphone. Most masks are the oronasal type, which covers only the mouth and nose. Large transport aircraft are usually fitted with smoke masks for each crew position. The smoke masks are installed in stowage containers within easy grasp of the individual. These masks provide crew protection in an emergency and are not used frequently like the demand and continuous flow masks.

[19] Passenger masks may be simple, cup shaped rubber moldings sufficiently flexible to obviate individual fitting. They may have a simple elastic head strap or they

may be held to the face by the passenger.

NEW WORDS AND EXPRESSIONS

(1) atmosphere[ˈætməsfɪə]　　　　大气
(2) oxygen[ˈɔksɪdʒən]　　　　氧气
(3) lung[lʌŋ]　　　　肺
(4) portable[ˈpɔːtəbl]　　　　手提式
(5) chemical[ˈkemɪkəl]　　　　化学的
(6) solid[ˈsɔlɪd]　　　　固态的；固体
(7) disadvantage[ˌdɪsədˈvɑːntɪdʒ]　　　　缺点
(8) generator[ˈdʒenəreɪtə]　　　　发生器，发电机
(9) sodium chlorate　　　　氯酸钠
(10) formula[ˈfɔːmjʊlə]　　　　化学式
(11) ingredient[ɪnˈgriːdjənt]　　　　成分
(12) oxygen candle　　　　氧烛
(13) packing[ˈpækɪŋ]　　　　包装
(14) squib[skwɪb]　　　　爆炸帽,电爆管
(15) insulation[ˌɪnsəˈleɪʃən]　　　　绝缘
(16) check valve　　　　单向阀
(17) relief valve　　　　释压阀
(18) proportional[prəˈpɔːʃənəl]　　　　成比例的
(19) barium peroxide　　　　过氧化钡
(20) barium dioxide　　　　二氧化钡
(21) alkaline[ˈælkəlɪn]　　　　碱性的
(22) mask[mɑːsk]　　　　面罩
(23) socket[ˈsɔkɪt]　　　　插座,外接口
(24) aneroid[ˈænərɔɪd]　　　　膜盒,无液的
(25) shutoff valve　　　　关断活门
(26) stainless steel　　　　不锈钢
(27) nonshatterable　　　　非易碎的
(28) dilution[daɪˈluːʃən]　　　　稀释

EXERCISE

1. **Translate the followings into Chinese.**
(1) **Concentration**. The concentration of a substance is the proportion of essential ingredients or substances in it.

(2) **Chemical oxygen generator**. A chemical oxygen generator is a device releasing oxygen created by a chemical reaction. The oxygen source is usually an inorganic superoxide, chlorate, or perchlorate. A promising group of oxygen sources are ozonides. The generators are usually ignited mechanically by a firing pin.

(3) **Sodium chlorate**. a colorless salt ($NaClO_3$) used as a weed killer and an antiseptic.

(4) **Oxygen candle**. An oxygen candle is a cylindrical chemical oxygen generator containing a mix of sodium chlorate and iron powder. When ignited, the mixture smolders at about 600℃ (1,112℉), producing sodium chloride, iron oxide, and about 6.5 man-hours of oxygen per kilogram of the mixture. It releases oxygen at a fixed rate.

(5) **Squib**. An electrically operated pyrotechnic device for breaking a disc in the fire extinguisher bottle to release the extinguishing fluid. Also used for escape slide deployment.

(6) **Chlorine**. Chlorine is a strong-smelling gas that is used to clean water and to make cleaning products.

(7) **Barium peroxide**. A white toxic powder obtained by heating barium oxide in air.

(8) **Orifice**. A specified size hole in component for control of fluid flow from or to the component.

(9) **Diluter demand regulator**. A component in the crew oxygen system which mixes air and oxygen and supplies constant oxygen pressure even if there are changes in ambient cockpit pressure.

(10) **Aneroid**. A sealed flexible device containing air at low pressure, which expands or contracts during change of pressure outside the device.

2. Answer the following questions in English.
(1) What are the two independent oxygen systems on Boeing airplane?
(2) What are the three forms of oxygen commonly used on aircraft?
(3) How to distinguish between high pressure oxygen cylinder and low pressure oxygen cylinder?
(4) How does the diluter demand regulator work?
(5) What are the three manual control levers on the regulator face?

SUPPLEMENTARY READING

[1] The servicing procedures for a gaseous oxygen system depend upon the type of system. Before charging an aircraft oxygen system, consult the aircraft manufacturer's maintenance manual. Precautions such as purging the connecting hose before coupling to

the aircraft filler valve, avoiding overheating caused by too rapid filling, opening cylinder valves slowly, and checking pressure frequently during charging should be considered.

[2] The type of oxygen to be used, the safety precautions, the equipment to be used, and the procedures for filling and testing the system must be observed. Gaseous breathing oxygen used in aircraft is a special type of oxygen containing practically no water vapor and is at least 99.5 percent pure. While other types of oxygen (welder, hospital) may be not pure enough, they usually contain water, which might freeze and block the oxygen system plumbing especially at high altitudes.

[3] Gaseous breathing oxygen is generally supplied in 220 to 250 cubic feet high pressure cylinders. The cylinders are identified by their dark green color with a white band painted around the upper part of the cylinder. The words "OXYGEN AVIATORS' BREATHING" are also stenciled in white letters, lengthwise along the cylinders.

[4] Gaseous oxygen is dangerous and must be handled properly. It causes flammable materials to burn violently or even to explode. Listed below are several precautionary measures to follow:

- Tag all reparable cylinders that have leaky valves or plugs.
- Don't use gaseous oxygen to dust off clothing, etc.
- Keep oil and grease away from oxygen equipment.
- Don't service oxygen systems in a hangar because of the increased chances for fire.
- Valves of an oxygen system or cylinder should not be opened when a flame, electricalarc, or any other source of ignition is in the immediate area.
- Properly secure all oxygen cylinders when they are in use.

[5] Even though several types of servicing trailers are in use, each recharging system contains supply cylinders, various types of valves, and a manifold that connects the high pressure cylinders to a purifier assembly. In the purifier assembly, moisture is removed from the oxygen. Coarse particles are trapped in the filter before reaching a reducing valve. The reducing valve has two gauges which are used to monitor inlet and outlet pressures respectively. The reducing valve also has an adjusting screw for regulating the outlet pressure.

[6] On many aircraft a chart is located adjacent to the filler valve which shows the safe maximum charging pressure for the ambient temperature. This must be observed when charging the system. It is common practice to have a warning placard cautioning against using oil or grease on the filler connections. Oxygen ground equipment should be maintained to a standard of cleanliness comparable to that of the aircraft system.

[7] When it is necessary to drain the system, it can be done by inserting a filler adapter into the filler valve and opening the shutoff valves. Do not drain the system too

rapidly as this will cause condensation within the system. An alternate method of draining the system is opening the emergency valve on the demand oxygen regulator. Perform this job in a well ventilated area and observe all fire precautions.

[8] Always keep the external surfaces of the components of the oxygen system, such as lines, connections, and mounting brackets, clean and free of corrosion and contamination with oil or grease. As a cleaning agent, use anhydrous (waterless) ethyl alcohol, isopropyl alcohol (anti-icing fluid), or any other approved cleaner.

[9] If mask to regulator hoses are contaminated with oil or grease, the hoses should be replaced. Many materials, particularly oils, grease, and nonmetallic materials, are likely to burn when exposed to oxygen under pressure. To avoid fire or an explosion it is essential that all oxygen equipment be kept clean and free from oil or grease.

[10] An oxygen fire or explosion depends on a combination of oxygen, a combustible material, and heat. The danger of ignition is in direct ratio to the concentration of oxygen, the combustible nature of the material exposed to the oxygen, and the temperature of the oxygen and material. Oxygen itself does not burn but it supports and intensifies a fire with any combustible material.

[11] When working on an oxygen system it is essential that the warnings and precautions given in the aircraft maintenance manual be carefully observed. In general, before any work is attempted on an oxygen system the following fire precautions should be taken:

- Provide adequate fire fighting equipment.
- Display "NO SMOKING" placards.
- Avoid checking aircraft radio or electrical systems.
- Keep all tools and oxygen servicing equipments free from oil or grease.

TOPIC 14：材料相关专业词汇

中文	英文	中文	英文
表面硬化	case-hardened	回火脆性	temper embrittlement
标距	gage length	焊接	weld
不锈钢	stainless steel	环氧树脂	epoxy
表面涂层	surface coating	红外热成像术	infrared thermography
薄板	thin sheet	划伤	lacerations
表面缺陷	areal defect	回路	loop
剥落	peeling	花键轴	splined shaft
比模量	specific modulus	回火马氏体	tempered martensite
比强度	specific strength	横向压缩	transverse compression
表面蚀坑	surface pit	卷曲	curling
表面磨损	surface wear	晶间失效	intergranular failure
脆性裂纹	brittle crack	键槽	keyway
擦伤	galling	局部撕裂	local tearing
冲击断裂	impact fracture	机械损伤	mechanical damage
冲击试验	impact test	尖锐的切口	sharp cut
淬火	quench	剪切裂纹	shear crack
冲压	stamping	极限强度	ultimate strength
残余应力	residual stress	抗拉强度	tensile strength
分层剥离	delamination	加工硬化	work hardening
断裂力学	fracture mechanics	基体金属	matrix metal
断裂韧性	fracture toughness	挤压	extrusion
镀铬	chromium plating	晶粒边界	qrain boundary
短纤维	discontinuous fiber	聚合物	polymer
镀锌钢	galvanized steel	浸润	wetting
钝化	passivation	抗剪强度	shear strength
腐蚀疲劳	corrosion fatigue	抗拉强度	tensile strength
腐蚀抑制剂	corrosion inhibitor	扩散区	diffuse zone
防锈剂	rust inhibition	裂纹尖端	crack tip
腐蚀产物	corrosion product	临界应力强度因子	critical stress intensity factor
缝隙腐蚀	pore corrosion	裂纹快速扩展	rapid crack propagation
粉末冶金	powder metallurgy	拉伸试样	tensile sample
铬酸	chromic acid	拉伸曲线	tensile curve
干燥剂	desiccant	拉伸试验	tensile test
共熔	eutectic meting	冷加工	cold work
惯性矩	moment of inertia	裂纹生长	crack growth
公制换算	metric conversion	滑移	slip
火焰切割	flame cutting	裂纹扩展	crack spreading
力学性能	mechanical properties	热循坏	thermal cycling
毛刺	fin, icicle	试样制备	sample preparation
磨削	grinding	蚀刻	etching
扭转过载	torsion overload	时效	aging
扭曲,变形	distortion	渗碳	carburization
黏合	adhesion	失效机理	failure mechanism
黏结	bonding	弯曲试验	bend test
凝固	solidification	弹性变形	elastic deformation
疲劳开裂	fatigue cracking	弹性模量	elastic modulus

疲劳极限	fatigue limit	退火状态	annealed condition
疲劳断裂	fatigue fracture	铁素体	ferrite
切口	notch	微观组织	microstructure
屈服强度	yield strength	微振腐蚀	fretting corrosion
氢脆性	hydrogen embrittlement	相变温度	transition temperature
切割	cutting	显微照片	micrograph
蠕变速率	greep rate	应变硬化	strain hardening
热处理	heat treatment	应力强度	stress intensity
软毛刷	soft brush	预加压应力	compressive prestressing
热应力	thermal stress	应力集中	stress concentration
韧性	ductility	应力腐蚀	stress corrosion
润滑	lubricant	最大正应力	maximum normal stress
热膨胀系数	coefficient of thermal expansion	载荷循环	loading cycle
溶剂	flux	皱折	buckle
热压	hot pressing	钻孔	drilling
固熔处理	solution treating	轧制	rolling

UNIT 15

FIRE PROTECTION SYSTEM

[1] Because fire is one of the most dangerous threats to an aircraft, the potential fire zones of modern multi-engine aircraft are protected by a fixed fire protection system. A "fire zone" is an area or region of an aircraft designed by the manufacturer to require fire detection or fire extinguishing equipment and a high degree of inherent fire resistance. The term "fixed" describes a permanently installed system in contrast to any type of portable fire extinguishing equipment, such as a hand-held CO_2 fire extinguisher.

[2] A complete fire protection system on modern aircraft and on many older model aircraft includes both a fire detection system and a fire extinguishing system. The following list of detection methods includes those most commonly used in turbine engine aircraft fire protection systems. The complete aircraft fire protection system of most large turbine engine aircraft will incorporate several of these different detection methods: ①Rate-of-temperature-rise detectors; ②Radiation sensing detectors; ③Smoke detectors; ④Overheat detectors; ⑤Carbon monoxide detectors; ⑥ Combustible mixture detectors; ⑦Fiber-optic detectors; ⑧Observation of crew or passengers.

[3] A fire detection system should signal the presence of a fire. Units of the system are installed in locations where there are greater possibilities of a fire. Three detector systems in common use are the thermal switch system, thermocouple system, and the continuous-loop detector system.

[4] **A thermal switch system**: consists of one or more lights energized by the aircraft power system and thermal switches that control operation of the light(s). These thermal switches are heat-sensitive units that complete electrical circuits at a certain temperature. They are connected in parallel with each other but in series with the indicator lights (Fig.15-1). If the temperature rises above a set value in anyone section of the circuit, the thermal switch will close, completing the light circuit to indicate the presence of a fire or overheat condition. Some warning lights are the "push-to-test" type. The bulb is tested by pushing it in to complete an auxiliary test circuit. The circuit in Fig.15-1 includes a test relay. With the relay contact in the position shown, there are two possible paths for current flow from the switches to the light. This is an additional safety feature. Energizing the test relay completes a series circuit and checks all the

wiring and the light bulb. Also included in the circuit shown in Fig. 15-1 is a dimming relay. By energizing the dimming relay, the circuit is altered to include a resistor in series with the light. In some installations several circuits are wired through the dimming relay, and all the warning lights may be dimmed at the same time.

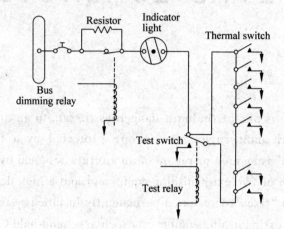

Fig. 15-1 Thermal switch system

[5] **The thermocouple fire warning system**: operates on an entirely different principle than the thermal switch system. A thermocouple depends upon the rate of temperature rise and will not give a warning when an engine slowly overheats or a short circuit develops. The system consists of a relay box, warning lights, and thermocouples. The wiring system of these units may be divided into the following circuits (Fig. 15-2): ①The detector circuit; ②the alarm circuit; ③the test circuit.

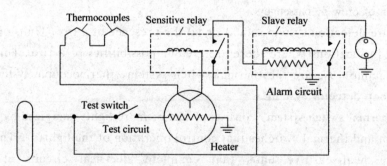

Fig. 15-2 Thermocouple system

[6] The relay box contains two relays, the sensitive relay and the slave relay. and the thermal test unit. Such a box may contain from one to eight identical circuits, depending on the number of potential fire zones. The relays control the warning lights. In turn, the thermocouples control the operation of the relays. The circuit consists of several thermocouples in series with each other and with the sensitive relay. The thermocouple is constructed of two dissimilar metals such as chrome and constantan. The point where these metals are joined and will be exposed to the heat of a fire is called a

hot junction. There is also a reference junction enclosed in a dead airspace between two insulation blocks. A metal cage surrounds the thermocouple to give mechanical protection without hindering the free movement of air to the hot junction. If the temperature rises rapidly, the thermocouple produces a voltage because of the temperature difference between the reference junction and the hot junction. If both junctions are heated at the same rate, no voltage will result and no warning signal is given. If there is a fire, however, the hot junction will heat more rapidly than the reference junction. The ensuing voltage causes a current to flow within the detector circuit. Any time the current is greater than 0.004 ampere, the sensitive relay will close. This will complete a circuit from the aircraft power system to the coil of the slave relay which closes and completes the circuit to the fire warning light.

[7] **A continuous-loop detector or sensing system** permits more complete coverage of a fire hazard area than any type of spot-type temperature detectors. Continuous-loop systems are versions of the thermal switch system. They are overheat systems, heat-sensitive units that complete electrical circuits at a certain temperature. There is no rate-of-heat-rise sensitivity in a continuous-loop system. Two widely used types of continuous-loop systems are the Kidde and the Fenwal systems.

[8] In the Kidde continuous-loop system (Fig. 15-3), two wires are imbedded in a special ceramic core within an Inconel tube. One of the two wires in the Kidde sensing system is welded to the case at each end and acts as an internal ground. The second wire is a hot lead (above ground potential) that provides a current signal when the ceramic core material changes its resistance with a change in temperature. Another continuous-loop system, the Fenwal system (Fig. 15-4), uses a single wire surrounded by a continuous string of ceramic beads in an Inconel tube. The beads in the Fenwal detector are wetted with a eutectic salt which possesses the characteristic of suddenly lowering its electrical resistance as the sensing element reaches its alarm temperature.

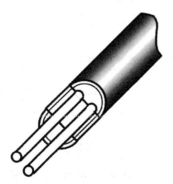 

Fig. 15-3　Kidde sensing element　　**Fig. 15-4　Fenwal sensing element**

[9] In both the Kidde and the Fenwal systems, the resistance of the ceramic or eutectic salt core material prevents electrical current from flowing at normal

temperatures. In case of a fire or overheat condition, the core resistance drops and current flows between the signal wire and ground, energizing the alarm system. The Kidde sensing elements are connected to a relay control unit. This unit constantly measures the total resistance of the full sensing loop. The system senses the average temperature, as well as any single hot spot. The Fenwal system uses a magnetic amplifier control unit. This system is non-averaging but will sound an alarm when any portion of its sensing element reaches the alarm temperature. Both systems continuously monitor temperatures in the affected compartments and both will automatically reset following a fire or overheat alarm after the overheat condition is removed or the fire extinguished.

[10] **The Lindberg fire detection system** (Fig. 15-5) is a continuous-element type detector consisting of a stainless steel tube containing a discrete element. This element has been processed to absorb gas in proportion to the operating temperature set point. When the temperature rises (due to a fire or overheat condition) to the operating temperature set point, the heat generated causes the gas to be released from the element. Release of the gas causes the pressure in the stainless steel tube to increase. This pressure rise mechanically actuates the diaphragm switch in the responder unit, activating the warning lights and an alarm bell. A fire test switch is used to heat the sensors, expanding the trapped gas. The pressure generated closes the diaphragm switch, activating the warning system.

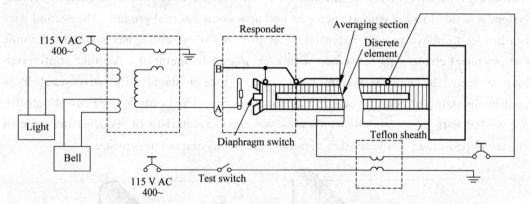

Fig. 15-5 Lindberg fire detection system schematic

[11] **Fire extinguisher system**: The typical fire extinguishing portion of a complete fire protection system includes a cylinder or container of extinguishing agent for each engine and nacelle area (Fig. 15-6). One type of installation provides for a container in each of four pylons on a multi-engine aircraft. This type of container is equipped with two discharge valves which are operated by electrically discharged cartridges. These two valves are the main and the reserve controls which release and route the agent to the pod and pylon in which the container is located or to the other engine on the same wing. This type of two-shot, cross feed configuration permits the release of a second charge of fire extinguishing agent to the same engine if another fire breaks out, without providing

two containers for each engine area.

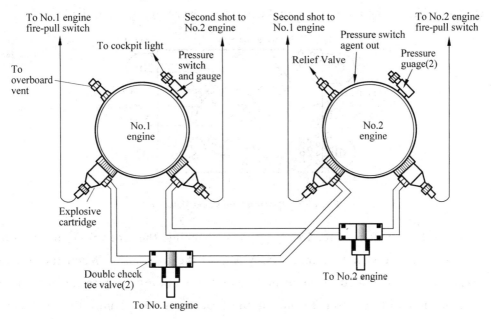

Fig. 15-6 Fire extinguisher system of a multi-engine aircraft

[12] A fire is detected in an engine, the appropriate fire switch handle light comes on. A solenoid energizes, releasing a mechanical interlock on the handle shaft. The handle is then operable. When the engine fire switch handle is pulled, the extinguishing bottles are armed for that engine, the engine generator field relay and generator circuit breaker are tripped, the engine fuel shut-off valve closes, the engine hydraulic shut-off valve closes, the bleed air valve closes, power to the engine thrust reverser isolation valve is removed, and the hydraulic warning lights illuminate. After pulling a fire switch handle, rotating it discharges the extinguishing agent into the appropriate engine. Rotating the handle counterclockwise discharges left bottle. Rotating it clockwise discharges right bottle.

[13] **Smoke detection system**: A smoke detection system monitors the cargo and baggage compartments for the presence of smoke, which is indicative of a fire condition. Smoke detection instruments, which collect air for sampling, are mounted in the compartments in strategic locations. A smoke detection system is used where the type of fire anticipated is expected to generate a substantial amount of smoke before temperature changes are sufficient to actuate a heat detection system.

[14] Typical smoke detection system is photoelectric smoke detector. This type of detector consists of a photoelectric cell, a beacon lamp, a test lamp, and a light trap (raster), all mounted on a labyrinth. An accumulation of 27% smoke in the air causes the photoelectric cell to conduct electric current. Fig. 15-7 shows the details of the smoke detector, and indicates how the smoke particles reflect the light to the

photoelectric cell. When activated by smoke, the detector supplies a signal to the smoke detector amplifier. The amplifier signal activates a warning light and bell.

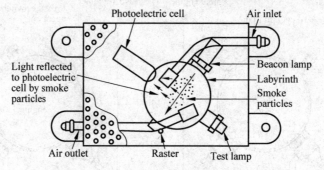

Fig. 15-7　Photoelectric smoke detector

[15] A test switch (Fig. 15-8) permits checking the operation of the smoke detector. Closing the switch connects 28 V DC to the test relay. When the test relay energizes, voltage is applied through the beacon lamp and test lamp in series to ground. A fire indication will be observed only if the beacon and test lamp, the photoelectric cell, the smoke detector amplifier, and associated circuits are operable.

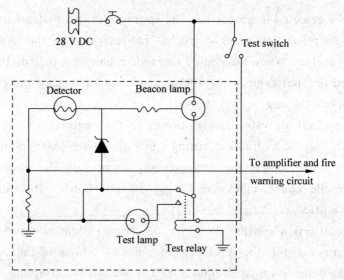

Fig. 15-8　Smoke detector test circuit

NEW WORDS AND EXPRESSIONS

(1) detection[dɪ'tekʃən]　　　　　　　　　　探测

(2) extinguish[ɪks'tɪŋwɪʃ]　　　　　　　　　熄灭

(3) extinguisher[ɪk'stɪŋwɪʃə]　　　　　　　灭火器

(4) rate-of-temperature-rise 温升速率
(5) detector [dɪˈtektə] 探测器
(6) radiation [ˌreɪdɪˈeɪʃən] 热辐射
(7) overheat [ˌəʊvəˈhiːt] 过热,使过热
(8) carbon monoxide 一氧化碳
(9) combustible [kəmˈbʌstəbəl] 可燃的
(10) fiber-optic 光学纤维的
(11) thermocouple [ˈθɜːməʊˌkʌpl] 热电偶
(12) continuous-loop 连续环路
(13) heat-sensitive 热敏感的
(14) dimming [ˈdɪmɪŋ] 变暗的
(15) chrome [krəʊm] 铬,铬合金
(16) constantan [ˈkɔnstəntæn] 康铜
(17) hot junction 热接点,热端
(18) hinder [ˈhɪndə] 阻碍
(19) ampere [ˈæmˌpɪə] 安培
(20) slave [sleɪv] 随动
(21) alarm [əˈlɑːm] 警报
(22) Kidde 基德
(23) Fenwal 芬沃尔
(24) inconel [ɪnˈkɔnel] 铬镍铁合金,因康镍合金
(25) eutectic [juːˈtektɪk] 共晶,共晶的
(26) Lindberg 林德伯格
(27) stainless [ˈsteɪnlɪs] 不锈钢的
(28) discrete [dɪsˈkriːt] 不连续的,离散的
(29) agent [ˈeɪdʒnt] 介质;剂
(30) armed [ɑːrmd] 预位,准备发射
(31) pod [pɔd] 容器,吊舱
(32) pylon [ˈpaɪˌlɔn] 吊挂
(33) crossfeed [krɔsfiːd] 交输
(34) indicative [ɪnˈdɪkətɪv] 指示的
(35) sampling [ˈsæmplɪŋ] 取样、采样
(36) photoelectric [ˌfəʊtəʊɪˈlektrɪk] 光电的
(37) labyrinth [ˈlæbərɪnθ] 集气室
(38) beacon [ˈbiːkən] 信标
(39) raster [ˈræstər] 光栅

EXERCISE

1. Translate the followings into Chinese.

(1) **Thermal radiation**. Thermal radiation is electromagnetic radiation generated by the thermal motion of charged particles in matter. All matter with a temperature greater than absolute zero emits thermal radiation. The mechanism is that bodies with a temperature above absolute zero have atoms or molecules with kinetic energies which are changing, and these changes result in charge-acceleration and/or dipole oscillation of the charges that compose the atoms. This motion of charges produces electromagnetic radiation in the usual way.

(2) **Optical fiber**. An optical fiber is a flexible, transparent fiber made of glass (silica) or plastic, slightly thicker than a human hair. It can function as a waveguide, or "light pipe". Optical fibers are widely used in fiber-optic communications, which permits transmission over longer distances and at higher bandwidths (data rates) than other forms of communication. Fibers are also used for illumination, and are wrapped in bundles so that they may be used to carry images, thus allowing viewing in confined spaces. Specially designed fibers are used for a variety of other applications, including sensors and fiber lasers.

(3) **Thermocouple**. A thermocouple consists of two dissimilar conductors in contact, which produce a voltage when heated. The size of the voltage is dependent on the difference of temperature of the junction to other parts of the circuit.

(4) **Constantan**. Constantan is a copper-nickel alloy usually consisting of 55% copper and 45% nickel. Its main feature is its resistivity which is constant over a wide range of temperatures. Other alloys with similarly low temperature coefficients are known, such as manganin ($Cu_{86}Mn_{12}Ni_2$).

(5) **Inconel alloys**. Inconel alloys are typically used in high temperature applications. Inconel alloys are oxidation and corrosion resistant materials well suited for service in extreme environments subjected to pressure and heat. When heated, Inconel forms a thick, stable, passivating oxide layer protecting the surface from further attack.

(6) **Eutectic system**. A eutectic system is a mixture of chemical compounds or elements that has a single chemical composition that solidifies at a lower temperature than any other composition made up of the same ingredients.

(7) **Fire extinguisher**. A fire extinguisher, flame extinguisher, or simply an extinguisher, is an active fire protection device used to extinguish or control small fires, often in emergency situations. It is not intended for use on an out-

of-control fire, such as one which has reached the ceiling, endangers the user (i.e., no escape route, smoke, explosion hazard, etc.), or otherwise requires the expertise of a fire department. Typically, a fire extinguisher consists of a hand-held cylindrical pressure vessel containing an agent which can be discharged to extinguish a fire.

(8) **Pylon**. The structure which holds an engine nacelle or pod to the wing or fuselage of a jet-propelled aircraft.

(9) **Sampling**. Taking a representative portion of a material or product to test (e.g. by physical measurements, chemical analysis, microbiological examination), typically for the purposes of identification, quality control, or regulatory assessment.

(10) **Beacon**. A beacon is an intentionally conspicuous device designed to attract attention to a specific location. Beacons can also be combined with semaphoric or other indicators to provide important information, such as the status of an airport, by the colour and rotational pattern of its airport beacon, or of pending weather as indicated on a weather beacon mounted at the top of a tall building or similar site. When used in such fashion, beacons can be considered a form of optical telegraphy.

2. Answer the following questions in English.
(1) What does a complete fire protection system constitute on modern aircraft?
(2) What are the detection methods of fire detection system?
(3) What are the three types of detectors most commonly used for fast detection of fires?
(4) How does the Lindberg fire detection system work?
(5) What is the working principle of a photoelectric smoke detector?

SUPPLEMENTARY READING

[1] Fire detector sensing elements are located in many high activity areas around aircraft engines. Their location, together with their small size, increases the chances of damage to the sensing elements during maintenance. The installation of the sensing elements inside the aircraft cowl panels provides some measure of protection not afforded elements attached directly to the engine. On the other hand, the removal and reinstallation of cowl panels can easily cause abrasion or structural defects to the elements. A well rounded inspection and maintenance program for all types of continuous loop systems should include the following visual checks. These procedures are provided as examples and should not be used to replace approved local maintenance directives or the applicable manufacturer's instructions. Sensing elements should be

inspected for:
(1) Cracked or broken sections caused by crushing or squeezing between inspection plates, cowl panels, or engine components.
(2) Abrasion caused by rubbing of element on cowling, accessories, or structural members.
(3) Pieces of safety wire or other metal particles which may short the spot detector terminals.
(4) Condition of rubber grommets in mounting clamps, which may be softened from exposure to oils, or hardened from excessive heat.
(5) Dents and kinks in sensing element sections. Limits on the element diameter, acceptable dents or kinks, and degree of smoothness of tubing contour are specified by manufacturers. No attempt should be made to straighten any acceptable dent or kink, since stresses may be set up that could cause tubing failure.
(6) Loose nuts or broken safety wire at the end of the sensing elements. Loose nuts should be retorqued to the value specified in the manufacturer's instructions. Some types of sensing element connections require the use of copper crush gaskets. These gaskets should be replaced any time a connection is separated.
(7) Broken or frayed flexible leads, if used. The flexible lead is made up of many fine metal strands woven into a protective covering surrounding the inner insulated wire. Continuous bending of the cable or rough treatment can break these fine wires, especially those near the connectors. Broken strands can also protrude into the insulated gasket and short the center electrode.
(8) Proper sensing element routing and clamping. Long unsupported sections may permit excessive vibration which can cause breakage. The distance between clamps on straight runs is usually about 8 to 10 inches, and is specified by each manufacturer. At end connectors, the first support clamp is usually located about 4 to 6 inches from the end connector fittings. In most cases, a straight run of 1 inch is maintained from all connectors before a bend is started, and an optimum bend radius of 3 inches is normally adhered to.
(9) Rubbing between a cowl brace and a sensing element. This interference, in combination with loose rivets holding the clamps to skin, may cause wear and short the sensing element.
(10) Correct grommet installation. The grommets are installed on the sensing element to prevent the element from chafing on the clamp. The slit end of the grommet should face the outside of the nearest bend. Clamps and grommets should fit the element snugly.
(11) Thermocouple detector mounting brackets should be repaired or replaced when cracked, corroded, or damaged. When replacing a thermocouple detector,

note which wire is connected to the identified plus terminal of the defective unit and connect the replacement in the same way.

(12) Test the fire detection system for proper operation by turning on the power supply and placing the fire detection test switch in the "TEST" position. The red warning light should flash on within the time period established for the system. On some aircraft an audible alarm will also sound.

[2] In addition, the fire detection circuits are checked for specified resistance and for an open or grounded condition. Tests required after repair of replacement of units in a fire detection system or when the system is inoperative include: (1) Checking the polarity, ground, resistance and continuity of systems that use thermocouple detector units, and (2) resistance and continuity tests performed on systems with sensing elements or cable detector units. In all situations follow the recommended practices and procedures of the manufacturer of the type system with which you are working.

[3] The following troubleshooting procedures represent the most common difficulties encountered in engine fire detection systems.

(1) Intermittent alarms are most often caused by an intermittent short in the detector system wiring. Such shorts may be caused by a loose wire which occasionally touches a nearby terminal, a frayed wire brushing against a structure, or a sensing element rubbing long enough against a structural member to wear through the insulation. Intermittent faults can often best be located by moving wires to recreate the short.

(2) Fire alarms and warning lights can occur when no engine fire or overheat condition exists. Such false alarms can most easily be located by disconnecting the engine sensing loop from the aircraft wiring. If the false alarm continues, a short must exist between the loop connections and the control unit. If, however, the false alarm ceases when the engine sensing loop is disconnected, the fault is in the disconnected sensing loop, which should be examined for areas which have been bent into contact with hot parts of the engine. If no bent element can be found, the shorted section can be located by isolating and disconnecting elements consecutively around the entire loop.

(3) Kinks and sharp bends in the sensing element can cause an internal wire to short intermittently to the outer tubing. The fault can be located by checking the sensing element with a megger while tapping the element in the suspected areas to produce the short.

(4) Moisture in the detection system seldom causes a false fire alarm. If, however, moisture does cause an alarm, the warning will persist until the contamination is removed or boils away and the resistance of the loop returns to its normal value.

(5) Failure to obtain an alarm signal when the test switch is actuated may be caused

by a defective test switch or control unit, the lack of electrical power, inoperative indicator light, or an opening in the sensing element or connecting wiring. When the test switch fails to provide an alarm, the continuity of a two wire sensing loop can be determined by opening the loop and measuring the resistance. In a single wire, continuous loop system, the center conductor should be grounded.

TOPIC 15：飞机缺陷常用术语

1. 结构缺陷

凹槽(gouge)：由于与另一相对锐利的物体接触而导致材料横截面改变，损伤区域形成一道连续且锋利或平滑的沟槽。

凹坑(dent)：在材料表面被挤压的损伤区域，其材料的横截面不改变，损伤的边缘平滑。此类损伤通常由另一表面平滑的物体撞击引起。凹坑的长度是凹坑边缘距离最长的两端的测量值，其宽度是贯穿凹坑的、垂直于长度的最大测量值。

变形(deform)：原有的状态发生变化。

擦伤(chafe)：因为部件间的过度摩擦导致损坏。

槽口/凹槽(notch)：V字形的切口或者突起。

穿孔/扎穿(puncture)：损伤贯穿整个材料的厚度，且无规则的形状，由另一尖锐物体的撞击或刺入而导致。

粗糙(rough)：表面不平整、隆起或突起的，不平滑的。

点腐蚀/沙眼(pitting)：表面较小的空穴，凹陷或凹痕。

断裂(fracture)：被打碎或打裂的状态。

分层(delaminate)：邻近两层材料之间因脱胶导致分离。

腐蚀(corrode)：由于综合的电化学反应导致的损伤，材料的横截面被改变。此类损伤多见于结构件的表面、孔壁或边缘等。

鼓起(bulge)：从原有形状表面向外凸起。

刮伤(scratch)：表面的薄而浅的切口或痕迹。材料上的任何深度或长度的切口均导致其横截面的改变。此类损伤通常因物体与另一极尖锐的物体接触导致。

积/含水(entrapped water)：水或水汽在某区域聚集。

刻痕/划痕(nick)：表面较浅的刻槽、刻痕或缺口。

裂纹/龟裂(crack)：材料内部因部分或全部断裂导致其横截面明显改变。此类损伤通常因振动、过载或内应力引起，其断裂往往不规则。

毛刺(burr)：被切割或钻孔后留在金属或其他材料上的粗糙的边缘。

磨蚀(fretting)：由很小的角位移或线性位移导致部件联接表面损伤，通常产生细小的黑色粉末。

磨损(abrasion)：由于刮、擦、划或其他表面侵蚀而导致的任何尺寸的表面损伤。此类损伤通常表面粗糙并且形状不规则。

扭曲(distort)：由于扭转、弯曲或持久的应力而导致的外形改变。此类损伤可由另一物体的撞击引起，但通常由邻近安装的部件的振动或位移影响而导致扭曲。

漆层剥落(finish missing(pertain to paint))：用于保护表面材料的层或保护层因失去黏附力而剥掉或脱落。

褪色/变色(fade)：失去原有的光泽或颜色。

下陷(sag)：因压力或策略而下沉、下垂或凹陷。

2. 金属材料缺陷

碎屑(chip)：从表面剥落或丢失的小碎片。

风蚀/风化(erode)：因气候或外来颗粒等原因导致表面被磨蚀。

霉/发霉(mildew)：有机质滋生霉菌而变质，常见白色的生物。

氧化(oxidize)：物质的原子失去电子的化学反应，是物质跟氧发生反应的过程，如金属的生锈。

翘曲(warp)：形状或外形轮廓扭曲变形。

3. 非金属材料缺陷

脱胶(debonding/disbond)：因为胶合失效导致材料的分离。

老化/恶化(deteriorate)：塑料等高分子化合物，在光、热、空气、机械力等的作用下，变得黏软或硬脆。

毛边/散股(fray)：因相磨或反复拉伸而使边缘破损、损耗。

撕裂(tear)：用力扯开或扯成碎片，撕碎。

受潮(wet)：湿的，沾满或浸透液体的(如水)。

磨损/陈旧(worn)：由于过度的磨耗而不可用。

起皱(wrinkle)：正常的平滑的表面上出现的皱纹或折痕。

4. 管路缺陷

爆裂(burst)：因为内部的压力过大导致(管路等)突然爆开或开裂。

污渍(stain)：外来物留下的污点或者污迹，使原物体变色。

错牙(cross-thread)：螺纹损坏导致相邻的螺牙相互交错。

渗漏(leak)：通过缺口或裂缝使某物(气体、液体等)逃逸、进入或通过。

走向错误(misrouted)：在行程中没有正确对齐。

堵塞(obstruct)：外物妨碍管道的畅通。

滑牙(stripped)：螺纹磨损、损坏至不可用状态。

扭结/弯曲(twist)：呈螺旋形或盘绕的形状，如两头转向相反的方向。

5. 电气线路缺陷

弯曲(bend)：改变平直或者原有的形状。

烧焦(burn)：被火、强热、辐射、电或腐蚀性物质损坏或伤害。

相碰(contact)：两个表面或物体的相碰或搭接。

脱开(disconnect)：使不连接，使分离断绝或中断两者的联系。

熔断(fused)：电路等因保险丝或电线烧断而断路。

间隙不够/没有间隙(insufficient/no clearance)：间隙测量值不足以满足要求。

扭结/缠结(kink)：在一段长度的材料产生的绷紧的卷曲、扭曲或弯曲。

短路(short)：因电路短接引起的故障。

6. 传动机构缺陷

游隙/无效行程(backlash)：由于齿轮或其他机械装置之间的连接处松动而造成的间隙。

卡阻(block)：运动部件的动作卡滞、不顺畅。
不一致(disagree)：对应关系不正确。
装错(wrong installation)：零部件没有安装在正确的位置。
咬死(seize)：由于高压或者高热使零件黏合在一起，导致运动困难。
松弛(钢索)(slack)：未拉紧的或未绷紧的，松散的。
迟滞(sluggish)：机械部件反应迟钝的表现。
干涩/缺乏润滑(un-lubricated)：因没有得到润滑而导致相磨、发热和磨损。

7. 紧固件缺陷
松动/松脱(loose)：未紧固的，未加限制的。
丢失(missing)：不在位，缺少的。
未关(扣、夹、卡)紧(not secure)：没有扣紧，不牢固或紧密。
剪断(shear)：因受剪切力而导致的脱离或丢失。
拖尾(trail)：移动或通过的某事物留下的记号或痕迹。

8. 其他部件或零件缺陷
不规则/不均匀(uneven)：不相等的，如在尺寸、长度、外表或质量上。
磨平(bald(= flat spot))：(轮胎)表面排水槽由于磨损而消失。
切口(cut)：指由另一锋利物体造成的裂口或伤口。
见线(轮胎表面)(exposed cords (pertain to tire))：轮胎编织线因磨损而导致裸露。
气压不足(low air pressure)：气体压力没有达到生产厂家的要求。
过期(expired)：超过时限、期限。
破损/缺损(damage)：对有用的或有价值的部件或设备的损伤或破坏。
模糊(not legible)：难以辨认的，无法读出或解释的。
被释放(电池/气瓶)(discharge)：使(蓄电池或气瓶等)所蓄的电或能量释放出来。
散包(滑梯等)(not packed)：(梯包)散开的状态。
变形/密封不良(滑梯等)(not sealed)：固定的包装被撕裂，露出里面的东西。

9. 一般性缺陷
摆动(fluctuate)：无规律连续的起伏不定的变化。
不工作(inoperative)：不能正常地发挥功能、作用。
不够/不足(insufficient)：达不到所要求的数量、质量、程度。（**注意，使用此术语用以描述缺陷或故障时必须列明要求的数据。**）
不亮(not illuminate)：不照明、不发光。
不能复位(will-not-reset)：不能回到原来的位置。
超标(exceed limit)：超出生产厂家的技术标准或规格要求的范围。
抖动(flutter)：快速的摇动或者上下振动。
断裂/断丝(broken)：被分成两半或更多块。
干扰(interfere)：妨碍无线电设备正常接收信号的电磁振荡。
漏装(not installed)：根据要求本应安装的零部件未安装。
失效(fail)：不能执行或不能完全执行原有的功能。（**注意，使用此术语用以描述缺陷**

或故障时必须列明要求的数据。)

污染(contaminate)：通过接触或混合使不纯净、不洁净。

相磨(rubbing)：不相关的物体表面之间互相接触(因相互间相对运动可能导致磨损)。

脏(dirty)：不干净，有灰尘或油渍。

折痕(crease)：因压弯或折叠导致弯折处出现明显的线痕或皱纹，一般认为折痕等同于裂纹。

UNIT 16

PNEUMATIC SYSTEM

[1] The purpose of the pneumatic system is to supply bleed air from the engine compressor or from the APU or from a ground cart. Air is distributed by a pneumatic manifold from the above sources to the air conditioning packs, wing and cowl thermal Anti-Ice (TAI) systems, the engine starting system, potable water system and the hydraulic reservoirs (Fig. 16-1). The pneumatic manifold interconnects the engine bleed system and APU and distributes bleed air to the systems. The pneumatic manifold contains the necessary valves to shut off bleed air at each engine, APU and to isolate the left and right hand systems.

[2] **On B737NG airplane**: The pneumatic manifold is separated into left and right sections by an isolation valve. A center section contains the APU duct. Flow of APU air into the center section is controlled by the APU bleed air valve. A ground service pneumatic connector is in the right section. Ground pneumatic sources bypass the pressure regulating valve and must have pressure regulation equipment.

[3] The main supply of bleed air to the manifold is obtained from the engine low pressure stage of high pressure compressor. Here air is supplied to the pneumatic manifold at all times except at engine idle or at low engine thrust settings. When this occurs high pressure stage bleed air is automatically used. Air from the bleed ports is controlled by the engine bleed control system. APU bleed air or pneumatic ground cart bleed air is primarily used for engine starting and for air conditioning pack operation on the ground. The APU can be used as an alternate bleed air source up to airplane altitude of 17,000 ft.

[4] Each engine bleed system contains a bleed air temperature control sub-system. This consists of a precooler, precooler control valve and precooler control valve sensor. The engine bleed air temperature is automatically controlled to a preset temperature whenever the engine is operating and the pressure regulator and shutoff valve (PRSOV) is open. The temperature is sensed in the precooler bleed air discharge. The sensor will cause the precooler control valve to modulate open/closed as required to maintain this temperature. Engine fan air is used as the heat sink for the bleed air.

[5] Two pressure transmitters are provided for pressure indication of the bleed air.

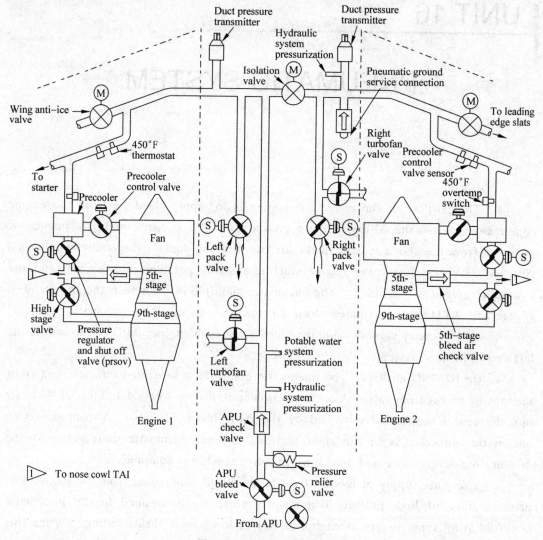

Fig. 16-1 Pneumatic system

One transmitter is used for each engine. Both pressure transmitters are connected to a dual pressure indicator on the overhead panel. The engine bleed air overtemperature indication consists of overtemperature switches in the ducting system, which are connected to the overhead panel trip lights. The temperature switches illuminate the trip lights and the corresponding engine PRSOV closes when the bleed air temperature exceeds approximately 490°F. Also, the bleed trip light will illuminate if there is excessive pressure at the bleed valve inlet.

[6] **On A320 airplane**: air is generally bled from an intermediate pressure (IP) stage of the engine high pressure (HP) compressor to minimize engine pressure losses. This is the normal engine air-bleed configuration:

- The IP stage is the 7th HP compressor stage. At low engine speeds, when the

pressure from the IP stage is insufficient, air is automatically bled from a higher compressor stage (HP stage). This happens especially at some aircraft holding points and during descent, with engines at idle.
- The HP stage is the 10th HP compressor stage. Transfer of air bleed is achieved by means of a pneumatically operated-butterfly valve, designated HP bleed valve.

[7] When the HP bleed valve is closed, air is directly bled from the IP stage through an IP bleed check valve, fitted with two flappers. When the HP bleed valve is open, the HP stage pressure is admitted into the pneumatic ducting and closes the IP bleed check valve. Air is then bled from the HP stage only. The HP bleed valve operates pneumatically and is connected by a sense line to a bleed pressure regulator valve and to a solenoid.

[8] The bleed pressure regulating valve is installed in the duct downstream of the IP bleed check valve and the HP bleed valve. The bleed pressure regulating valve also operates pneumatically but opening and closing can be controlled by a bleed pressure regulating valve control solenoid. The solenoid is installed in the duct downstream of the bleed air precooler exchanger. The solenoid controls the bleed pressure regulating valve which controls the HP bleed valve at the same time.

[9] **APU bleed system**: The aircraft has an APU bleed-air supply and a crossbleed system installed in the AFT and Mid fuselage. The load compressor of the APU supplies the APU bleed air to the distribution system. The supply of the APU bleed air is controlled by the APU bleed control valve. When the APU bleed control valve is in the open position, the pressure reducing valves (PRV) of the engines close, and shut off the engine bleed-air supply. Thus, the distribution system supplies the APU bleed-air to the user systems. It is not necessary to regulate the temperature, pressure and flow of the APU bleed-air because they agree with the user demand. The crossbleed system connects the left and right wing supply ducts. The left and right bleed-air systems work in correlation. A crossbleed valve isolates them from each other.

[10] Normally the APU check valve prevents engine bleed pressure from backing into the APU bleed system. Should the check valve fail open, with the APU bleed valve open, engine bleed pressure could be detrimental to the APU. The DUAL BLEED light on the overhead panel illuminates when the APU bleed valve is open and either Engine No. 1 PRSOV or Engine No. 2 PRSOV and isolation valve are open. The light would be on during an APU engine start. After engine start and the APU bleed valve switch is placed to "OFF", the light goes out indicating that the APU bleed valve has closed.

[11] **Ground cart bleed**: The external air supplies the air through a HP ground connector. This connector is installed on the left side of the belly fairing. The connector contains with a non-return valve and a built-in nipple. A duct connects the HP ground connector with the crossfeed duct. The joint of both ducts is on the left side of the

crossfeed valve.

[12] **PRSOV**: The PRSOV is a pneumatically actuated, spring loaded closed, butterfly type valve. The valve, in conjunction with a remotely mounted bleed air regulator, provides pneumatic signals to open, close, and regulate pressure for downstream bleed air. The valve consists basically of an actuator, manual override with closed position lock, and flow body with a butterfly plate.

[13] **Bleed air regulator**: The bleed air regulator is a pneumatic controller designed to provide regulated control pressure, with electrically controlled shutoff, from a bleed air source. Other functions incorporated are electrical indication of excessive bleed air supply pressure, automatic electrical shutoff of regulated control pressure in the event that downstream pressure exceeds bleed air supply pressure, and a relief valve to maintain regulation of control pressure in the event of pressure regulator failure.

[14] **Isolation valve**: The bleed air isolation valve is installed in the center of the crossover duct to separate the left and right pneumatic systems or to connect them when necessary. The isolation valve is a 115-volt, single-phase, motor operated butterfly valve. The valve normally stays closed unless a cross-feed of bleed air is necessary. The valve is controlled by a three-position (OPEN-AUTO-CLOSE) switch on the P5 panel. It is connected electrically through the left and right pack switches and also through the No. 1 and No. 2 engine bleed switches.

[15] **Precooler system**: The purpose of the bleed air precooler system is to control the temperature of the high pressure bleed air which is supplied to the manifold system to approximately 415°F. Two precooler systems are provided, one for engine No. 1 and one for engine No. 2. Each precooler system consists of an air-to-air heat exchanger, four-inch-diameter precooler control valve, a precooler control valve sensor and a solenoid valve installed in the sense line between the valve and sensor. The precooler is an air-to-air heat exchanger located at top right center of each engine and is of the plate-fin single pass crossflow type. The heat collected from the hot air ducted from the engine is absorbed by the cool air ducted from the engine fan after passing through the precooler control valve. The cooling air is then discharged overboard.

NEW WORDS AND EXPRESSIONS

(1) compressor[kəm'presə(r)]　　　压气机
(2) equipment['ɪkwɪpmənt]　　　　设备
(3) stage[steɪdʒ]　　　　　　　　　级
(4) idle['aɪdl]　　　　　　　　　　慢车
(5) setting['setɪŋ]　　　　　　　　设置
(6) precooler[priːˈkuːlə]　　　　　预冷器

(7) crossbleed[krɔsbliːd]　　　交输引气
(8) discharge[dɪsˈtʃɑːdʒ]　　　排放，放电
(9) sensor[ˈsensə(r)]　　　传感器
(10) transmitter[trænsˈmɪtə(r)]　　　传感器，发射机，发报机
(11) dual[ˈdjuːəl]　　　双的
(12) intermediate[ˌɪntəˈmiːdiət]　　　中间的
(13) belly[ˈbelɪ]　　　机腹
(14) flapper[ˈflæpə(r)]　　　舌门，活动挡板
(15) detrimental[ˌdetrɪˈmentl]　　　有损害的
(16) plate[pleɪt]　　　板
(17) nipple[ˈnɪpl]　　　螺纹接管，短管头
(18) joint[dʒɔɪnt]　　　接头
(19) butterfly[ˈbʌtəflaɪ]　　　蝶形
(20) override[ˌəʊvəˈraɪd]　　　超控
(21) diameter[daɪˈæmɪtə(r)]　　　直径

EXERCISES

1. Translate the followings into Chinese.

(1) **APU**. An APU is a device on a vehicle that provides energy for functions other than propulsion. They are commonly found on large aircraft, as well as some large land vehicles. Aircraft APUs generally produce 115 V at 400 Hz (rather than 50/60 Hz in mains supply), to run the electrical systems of the aircraft; others can produce 28 V DC. APUs are also present on naval ships. APUs can provide power through single or three-phase systems.

(2) **Compressor**. An air compressor is a device that converts power (usually from an electric motor, or an engine) into kinetic energy by compressing and pressurizing air, which, on command, can be released in quick bursts. There are numerous methods of air compression, divided into either positive-displacement or negative-displacement types.

(3) **Idle**. Idling is running a vehicle's engine when the vehicle is not in motion. When an engine idles, the engine runs without any loads except the engine accessories.

(4) **Transmitter**. A transmitter is a telemetry device which converts measurements from a sensor into a signal, and sends it, usually via wires, to be received by some display or control device located a distance away.

(5) **Bleed air**. Bleed air in gas turbine engines is compressed air taken from within the engine, after the compressor stage(s) and before the fuel is injected in the

burners. Bleed air is valuable in an aircraft for two properties: high temperature and high pressure (typical values are 200-250℃ and 275 kPa (40 PSI), for regulated bleed air exiting the engine pylon for use throughout the aircraft).

(6) **PRSOV**. Pressure regulator and shutoff valve (PRSOV) is a valve in the pneumatic system for regulating bleed air pressure from the engine to the pneumatic manifold.

(7) **Sensor**. An electrical device for detecting variable conditions and sending a signal to indicators and lights.

(8) **Butterfly valve**. A butterfly valve is a valve which can be used for isolating or regulating flow. The closing mechanism takes the form of a disk. Operation is similar to that of a ball valve, which allows for quick shut off. The disc is positioned in the center of the pipe, passing through the disc is a rod connected to an actuator on the outside of the valve. Rotating the actuator turns the disc either parallel or perpendicular to the flow.

(9) **Flapper valve**. Flapper valve is a swing check valve in which the disc, the movable part to block the flow, swings on a hinge or trunnion, either onto the seat to block reverse flow or off the seat to allow forward flow.

(10) **Isolation valve**. A valve in the system for separation of the system into sections.

2. **Answer the following questions in English.**
(1) What is the function of aircraft pneumatic system?
(2) What is the function of APU bleed check valve?
(3) What are the functions of PRSOV?
(4) What is the function of precooler?
(5) Does the pressure and temperature of APU bleed air need to regulate? Why?

SUPPLEMENTARY READING

[1] The pneumatic distribution system connects air supply sources from the APU, engines, and ground air source to user systems through the pneumatic manifold system and their appropriate control valves. The pneumatic manifold system extends from the engine at one wing to the crossover duct in the air conditioning bay below the center wing section to the other engine at the opposite wing. An electrically actuated isolation valve in the crossover duct separates the left and right side system. On B737 airplane, the APU bleed air duct is connected to the crossover duct on the left side of the isolation valve and the pneumatic ground service connection is connected to the crossover duct on the right side of the isolation valve.

[2] From the engine, bleed air source is ducted primarily from the 5th-stage port

where it passes through the 5th-stage check valve before it is ducted together with the 9th-stage (high pressure) duct. Air for nose cowl thermal anti-ice (TAI) is tapped off downstream of the 5th-stage check valve. Bleed air from the 9th-stage ports are ducted through the 9th-stage manifold and high stage valve before it is joined together with the 5th-stage duct. The air is then ducted through the PRSOV to regulate the pressure and the precooler heat exchanger to regulate the temperature before it is discharged into the strut duct. The strut duct is connected to the engine starter duct and to the wing leading edge duct. Air for wing thermal anti-ice (TAI) is tapped off from the wing leading edge duct through the wing thermal anti-ice valve. The left and right wing leading edge ducts are routed inboard inside the wing leading edge fairings where the ducts are joined to the crossover duct in the air conditioning bay.

[3] Bleed air source can also be supplied with use of the APU. The APU bleed air duct runs under the passenger floor from the APU along the left side of the aft cargo compartment, then inside the keel beam through the wheel well to the air conditioning bay where it joins the crossover duct on left side of the isolation valve. The APU bleed air (shutoff) valve, a pneumatically actuated solenoid controlled valve installed in the duct downstream of the APU, controls the airflow from the APU. A reverse flow APU check valve installed in the duct section downstream of the APU bleed air (shutoff) valve prevents damage to the APU compressor from bleed air backflows from the engine(s) or ground air source. A pressure relief valve is installed in the duct section between the APU bleed air valve and APU check valve to relief excess pressure in the pneumatic manifold.

[4] Three-inch diameter tap with an integral check valve for pneumatic ground service connection is installed on the right side of the crossover duct to allow system pressurization from an external ground air source.

[5] There are two duct pressure transmitters installed in the crossover duct to monitor duct pressure on either side of the isolation valve.

[6] The pneumatic manifold serves as the central reservoir for the supply of pressurized air for air conditioning, hydraulic system pressurization, and water tank pressurization through their applicable control valves from pressure taps in the pneumatic duct.

TOPIC 16：ICAO 简介

国际民用航空组织（International Civil Aviation Organization，ICAO）是联合国的一个专门机构，1944 年为促进全世界民用航空安全、有序的发展而成立，总部设在加拿大蒙特利尔，负责制定国际空运标准和条例，是 193 个缔约国（截至 2022 年）在民航领域中开展合作的媒介。

国际民航组织由大会、理事会和秘书处三级框架组成：

（1）大会：是国际民航组织的最高权力机构，由全体成员国组成。大会由理事会召集，一般情况下每三年举行一次，遇有特别情况或经五分之一以上成员国向秘书长提出要求，可以召开特别会议。大会决议一般以超过半数通过。参加大会的每一个成员国只有一票表决权。但在某些情况下，如《芝加哥公约》的任何修正案，则需三分之二多数票通过。

（2）理事会：是向大会负责的常设机构，由大会选出的 36 个缔约国组成。理事国分为三类：第一类是在航空运输领域居特别重要地位的成员国；第二类是对国际航空运输的发展有突出贡献的成员国；第三类是区域代表成员国。比例分配为 11：12：13，中国为第一类理事国。理事会每年召开三次会议，每次会议会期约为两个月。理事会下设财务、技术合作、非法干扰、航行、新航行系统、运输、联营导航、爱德华奖八个委员会。

（3）秘书处：是国际民航组织的常设行政机构，由秘书长负责保证国际民航组织各项工作的顺利进行，秘书长由理事会任命。秘书处下设航行局、航空运输局、法律局、技术合作局、行政局五个局以及财务处、外事处。此外，秘书处有一个地区事务处和七个地区办事处，分设在曼谷、开罗、达喀尔、利马、墨西哥城、内罗毕和巴黎。地区办事处直接由秘书长领导，主要任务是建立和帮助缔约各国实行国际民航组织制定的国际标准和建设措施以及地区规划。

国际民航组织的法规体系架构分为 4 个层级：《芝加哥公约》、《芝加哥公约》附件、空中航行服务程序（PANS）/地区补充程序（SUPP）、指导材料。其中附件包括 19 个：

附件 1：人员执照（personnel licensing），给飞行机组、空中交通管制员和航空器维修人员颁发执照。

附件 2：空中规则（rules of the air），航空器在飞行中或在机场活动区的运行必须遵守的一般规则。飞行中还必须遵守目视飞行规则和仪表飞行规则。

附件 3：国际航行的气象服务（meteorological service for international air navigation），向国际航行提供气象服务以报告从飞机上观察到的气象情报。

附件 4：航图（aeronautical charts），规定了航图的标准和建议措施。

附件 5：在空中和地面操作中使用的计量单位与度量制度（measurement units used in air and ground operations），该附件列出在空中和地面操作中使用的量纲系统。

附件 6：航空器的运行（operation of aircraft），确保在全世界相类似的运行中，高于规定的最低安全水平的标准：

第一部分：国际商业航空运输-飞机：适用于经批准从事国际商业航空运输运行的运营人实施的飞机运行。

第二部分：国际通用航空-飞机：适用于使用飞机从事的国际通用航空运行。

第三部分：国际飞行-直升机：适用于从事国际商用航空运输运行或国际通用航空运行的所有直升机，除非该标准和建议措施不适用于从事航空作业的直升机。

附件7：航空器的国籍和登记标志（aircraft nationality and registration marks），关于对航空器登记和识别的要求。

附件8：航空器的适航性（airworthiness of aircraft），根据统一的程序对航空器进行检查并颁发证书。

附件9：简化手续（facilitation），规定了简化手续的标准和建议措施。

附件10：航空电信（aeronautical telecommunication），第1卷规定通信设备和系统的标准，第2卷规定标准的通信程序。

附件11：空中交通服务（air traffic services），包括建立和使用空中交通管制、飞行情报和告警服务的信息。

附件12：搜寻与援救（search and rescue），提供搜寻与援救所需要的组织和使用相关设施与服务的信息。

附件13：航空器失事调查（aircraft accident investigation），对航空器事故的通知、调查和报告做了统一的规定。

附件14：机场（aerodromes），关于机场的设计和设备规范。

附件15：航行情报服务（aeronautical information services），包含收集和分发飞行所需的航行情报的方法。

附件16：环境保护（environmental protection），包含航空器噪声审定、噪声监测和供制定土地使用规划的噪声影响范围的规范（第1卷），以及有关航空器发动机排放物的规范（第2卷）。

附件17：安全保卫（security），保护国际民用航空免受非法干扰，规定了保护国际民用航空免受非法干扰的办法。

附件18：危险品的安全运输（the safe transport of dangerous goods by air），包括危险品的标识、包装和运输规范。

附件19：安全管理（safety management），规定了安全管理的标准和建议措施。本附件所载的标准和建议措施，必须适用于与航空器安全运行有关或直接支持其安全运行的安全管理职责。

UNIT 17

ICE AND RAIN PROTECTION

[1] Rain, snow, and ice are enemies of the airplanes' safety. The two types of ice encountered during flight are glaze ice and rime ice. Ice on an aircraft affects its performance and efficiency in many ways. Ice buildup increases drag and reduces lift. It causes destructive vibration, and hampers true instrument readings. Control surfaces become unbalanced or frozen. Fixed slots are filled and movable slots jammed. Radio reception is hampered and engine performance is affected. Ice or rain accumulation on the windshield interferes with vision. The ice and rain protection system protects the airplane and aids the flight crew when operating under ice and rain conditions.

[2] Ice can be detected visually, but most modern aircraft have one or more ice detector sensors that warn the flight crew of icing conditions. An annunciator light comes on to alert the flight crew. In some aircraft models, multiple ice detectors are used, and the ice detection system automatically turns on the wing anti-icing (WAI) systems when icing is detected.

[3] Several means to prevent or control ice formation are used in aircraft today: ①Heating surfaces using hot air; ②heating by electrical elements; ③breaking up ice formations, usually by inflatable boots; ④alcohol spray. A surface may be anti-iced either by keeping it dry by heating to a temperature that evaporates water upon impingement; or by heating the surface-just enough to prevent freezing, maintaining it running wet; or the surface may be deiced by allowing ice to form and then removing it. Ice prevention or elimination systems ensure safety of flight when icing conditions exist. Ice may be controlled on aircraft structure by the following methods:

(1) Leading edge of the wing-thermal pneumatic, thermal electric.

(2) Leading edges of vertical and horizontal stabilizers-thermal pneumatic, thermal electric.

(3) Windshields, windows, and radomes-thermal, chemical.

(4) Engine air inlets-thermal pneumatic.

(5) Pitot and static air data sensors-thermal electrical.

(6) Lavatory drains and portable water lines-thermal electrical.

[4] Thermal systems used for the purpose of preventing the formation of ice or for

deicing airfoil leading edges usually use heated air ducted spanwise along the inside of the leading edge of the airfoil and distributed around its inner surface. These thermal pneumatic anti-icing systems are used for wings, leading edge slats, horizontal and vertical stabilizers, engine inlets, and more. There are several sources of heated air, including hot air bled from the turbine compressor, engine exhaust heat exchangers, and ram air heated by a combustion heater.

[5] **Wing anti-ice system:** Thermal wing anti-ice (WAI or TAI) systems for business jet and large-transport category aircraft typically use hot air bled from the engine compressor. Relatively large amounts of very hot air can be bled from the compressor, providing a satisfactory source of anti-icing heat. The hot air is routed through ducting, manifolds, and valves to components that need to be anti-iced. Fig. 17-1 shows a typical WAI system schematic for a business jet. The bleed air is routed to each wing leading edge by an ejector in each wing inboard area. The ejector discharges the bleed air into piccolo tubes for distribution along the leading edge. Fresh ambient air is introduced into the wing leading edge by two flush-mounted ram air scoops in each wing leading edge, one at the wing root and one near the wingtip. The ejectors entrain ambient air, reduce the temperature of the bleed air, and increase the mass airflow in the piccolo tubes. The wing leading edge is constructed of two skin layers separated by a narrow passageway (Fig. 17-2). The air directed against the leading edge can only escape through the passageway.

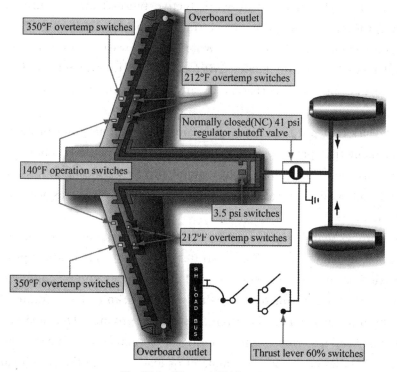

Fig. 17-1 Thermal WAI system

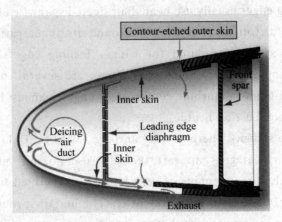

Fig. 17-2　Heated wing leading edge

[6] The WAI valve controls the flow of bleed air from the pneumatic system to the WAI ducts. The valve is electrically controlled and pneumatically actuated. The torque motor controls operation of the valve. With no electrical power to the torque motor, air pressure on one side of the actuator holds the valve closed. Electrical current through the torque motor allows air pressure to open the valve. As the torque motor current increases, the valve opening increases.

[7] **Thermal electric anti-icing**: Electricity is used to heat various components on an aircraft so that ice does not form. This type of anti-ice is typically limited to small components due to high amperage draw. Effective thermal electric anti-ice is used on most air data probes, such as pitot tubes, static air ports, TAT and AOA probes, ice detectors, and engine P2/T2 sensors. Water lines, waste water drains, and some turboprop inlet cowls are also heated with electricity to prevent ice from forming. Transport category and high-performance aircraft use thermal electric anti-icing in windshields.

[8] In devices that use thermal electric anti-ice, current flows through an integral conductive element that produces heat. The temperature of the component is elevated above the freezing point of water, so ice cannot form. Various schemes are used, such as an internal coil wire, externally wrapped blankets or tapes, as well as conductive films and heated gaskets.

[9] Data probes that protrude into the ambient airstream are particularly susceptible to ice formation in flight. Fig. 17-3 illustrates the types and location probes that use thermal electric heat on one airliner. A pitot tube, for example, contains an internal electric element that is controlled by a switch in the cockpit. Use caution checking the function of the pitot heat when the aircraft is on the ground. The tube gets extremely hot since it must keep ice from forming at altitude in temperatures near $-50°F$ at speeds possibly over 500 miles per hour. An ammeter or load meter in the circuit can be used as a substitute to touching the probe, if so equipped.

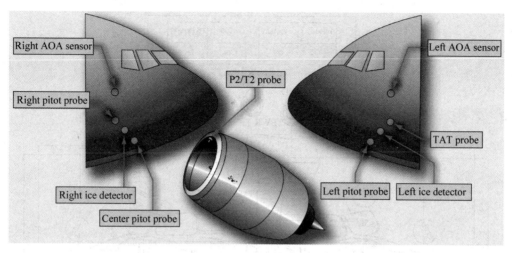

Fig. 17-3 Probes with thermal electric anti-icing

[10] **Windshield icing control systems**: In order to keep window areas free of ice, frost, etc., window anti-icing, deicing, defogging, and demisting systems are used. The systems vary according to the type of aircraft and its manufacturer. Some windshields are built with double panels having a space between, which will allow the circulation of heated air between the surfaces to control icing and fogging. Others use windshield wipers and anti-icing fluid which is sprayed on.

[11] One of the more common methods for controlling ice formation and fog on modern aircraft windows is the use of an electrical heating element built into the window. When this method is used with pressurized aircraft, a layer of tempered glass gives strength to withstand pressurization. A layer of transparent conductive material (stannic oxide) is the heating element and a layer of transparent vinyl plastic adds a non-shattering quality to the window. The vinyl and glass plies are bonded by the application of pressure and heat. The bond is achieved without the use of a cement as vinyl has a natural affinity for glass. The conductive coating dissipates static electricity from the windshield in addition to providing the heating element.

[12] An electrically heated windshield system includes: ① windshield autotransformers and heat control relays; ② heat control toggle switches; ③ indicating lights; ④ windshield control units; ⑤ temperature sensing elements (thermistors) laminated in the panel (Fig. 17-4).

[13] **Water and toilet drain heaters**: Heaters are provided for toilet drain lines, water lines, drain masts, and waste water drains when they are located in an area that is subjected to freezing temperatures in flight. The types of heaters used are integrally heated hoses, ribbon, blanket, or patch heaters that wrap around the lines, and gasket heaters. Thermostats are provided in heater circuits where excessive heating is undesirable or to reduce power consumption. The heaters have a low voltage output and continuous operation will not cause overheating.

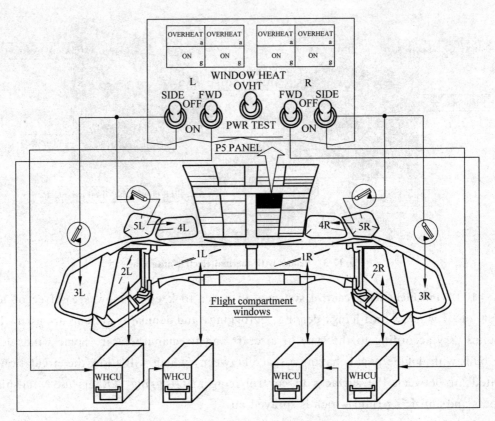

Fig. 17-4 Windows heat control units

[14] **Rain control systems**: When rain forms on a windshield during flight, it becomes a hazard and must be eliminated. To provide a clear windshield, rain is eliminated by wiping it off or blowing it off. There are several different ways to remove the rain from the windshields. Most aircraft use one or a combination of the following systems: windshield wipers, pneumatic rain removal, chemical rain repellent.

[15] **Windshield wipers**: Windshield wipers for aircraft are similar to those used on automobiles except they must be able to withstand the air loads caused by high speeds of operation. Electrical windshield wipers are usually operated by a two-speed DC motor that drives a converter. This converter changes the rotary output of the motor into the reciprocating motion needed for the wiper blades. When the windshield wiper switch is turned off, the control circuit is open, but the motor continues to run until the blades are driven to the Park position. The motor then stops, but the control circuit is armed so the motor will start when the windshield wiper switch is turned to either the fast or slow position. Some installations have a separate position on the speed selector switch that allows the pilot to drive the wiper blades to the park position before putting the switch in the "OFF" position (Fig. 17-5).

[16] Windshield wiper must never be operated on a dry windshield, and the blades must be kept clean and free of any type of contaminants that could scratch the

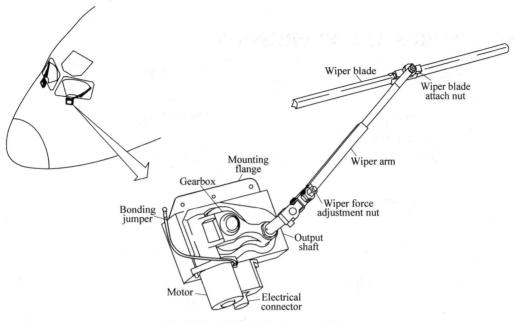

Fig. 17-5 Windshield wiper and drive

windshield. If the windshield wiper should ever have to be operated for maintenance or adjustment, the windshield must be flooded with ample quantities of fresh, clean water and kept wet while the wiper blades are moving across the glass.

[17] **Pneumatic rain removal systems**: Windshield wipers characteristically have two basic problem areas. One is the tendency of the slipstream aerodynamic forces to reduce the wiper blade loading pressure on the window, causing ineffective wiping or streaking. The other is in achieving fast enough wiper oscillation to keep up with high rain impingement rates during heavy rain falls. As a result, most aircraft wiper systems fail to provide satisfactory vision in heavy rain. With the advent of turbine powered aircraft, the pneumatic rain removal system became feasible. This method uses high pressure, high temperature engine compressor bleed air which is blown across the windshields. The air blast forms a barrier that prevents raindrops from striking the windshield surface.

[18] **Windshield rain repellant**: Many of the jet transport aircraft have a chemical rain repellent system that uses a liquid chemical sprayed on the windshield to prevent the water reaching the surface of the glass. Since it cannot wet the surface and spread out, the water will form beads and the wind can easily carry it away and leave the glass free of water so the pilot's visibility will not be distorted. The liquid should never be sprayed onto the windshield unless the rain is sufficiently heavy, because too much repellent can smear on the windshield and be difficult to see through. The repellent is difficult to remove if it is sprayed onto dry windshield.

NEW WORDS AND EXPRESSIONS

(1) protection[prəˈtekʃən] 保护，防护
(2) accumulate[əˈkjuːmjʊleɪt] 积聚，堆积
(3) buildup[ˈbɪldʌp] 积累，形成
(4) hamper[ˈhæmpə] 妨碍
(5) jam[dʒæm] 堵塞
(6) inflatable boot 膨胀管，膨胀带
(7) alcohol[ˈælkəˌhɒl] 酒精
(8) anti-ice 防冰
(9) deice[diːˈaɪs] 除冰
(10) heater[ˈhiːtə] 加热器
(11) windshield[ˈwɪndʃiːld] 风挡玻璃
(12) wiper[ˈwaɪpər] 雨刷
(13) stannic[ˈstænɪk] 锡的
(14) nonshattering[ˈnɒnʃætərɪŋ] 不易脆的
(15) vinyl plastic 乙烯基塑料
(16) cement[sɪˈment] 结合剂
(17) affinity[əˈfɪnɪtɪ] 亲和力
(18) autotransformer[ˌɔːtəʊtrænsˈfɔːmə] 自耦变压器
(19) laminate[ˈlæməˌneɪt] 分层，层压板
(20) thermostat[ˈθɜːməˌstæt] 恒温器
(21) repellant[rɪˈpelənt] 排雨剂
(22) converter[kənˈvɜːtə] 转换装置
(23) adjustment[əˈdʒʌstmənt] 调节，调整

EXERCISE

1. Translate the followings into Chinese.

(1) **Rime ice**. Rime ice is a white ice that forms when the water droplets in fog freeze to the outer surfaces of objects. It is often seen on trees atop mountains and ridges in winter, when low-hanging clouds cause freezing fog. This fog freezes to the windward (wind-facing) side of tree branches, buildings, or any other solid objects, usually with high wind velocities and air temperatures between −2℃ and −8℃ (28.4℉ and 17.6℉).

(2) **Glaze ice**. Glaze ice is a smooth, transparent and homogeneous ice coating occurring when freezing rain or drizzle hits a surface. It is similar in

appearance to clear ice, which forms from supercooled water droplets. It is a relatively common occurrence in temperate climates in the winter when precipitation forms in warm air aloft and falls into below-freezing temperature at the surface.

(3) **Radome**. A radome (the word is a contraction of radar and dome) is a structural, weatherproof enclosure that protects a microwave (e.g. radar) antenna. The radome is constructed of material that minimally attenuates the electromagnetic signal transmitted or received by the antenna.

(4) **Windshield**. A transparent glass or plastic covering used to protect the occupants of an aircraft from the wind and rain while allowing forward vision. A windshield is also called a wind-screen.

(5) **Anti-ice**. The prevention of the formation of ice on a surface. Ice may be prevented by the use of heat or by covering the surface with a chemical which prevents water adhering to the surface. Anti-ice should not be confused with deicing, which is the removal of ice after it has formed on the surface.

(6) **Pitot tube**. A pitot tube is a pressure measurement instrument used to measure fluid flow velocity. The pitot tube was invented by the French engineer Henri Pitot in the early 18th century and was modified to its modern form in the mid-19th century by French scientist Henry Darcy. It is widely used to determine the airspeed of an aircraft and to measure air and gas velocities in industrial applications. The pitot tube is used to measure the local velocity at a given point in the flow stream and not the average velocity in the pipe or conduit.

(7) **Autotransformer**. A single-winding electrical transformer that uses a carbon brush riding on a bare portion of the winding to select the number of turns used as the secondary. Autotransformers produce a secondary voltage that can be varied from almost zero to a value higher than the primary voltage. Autotransformers do not isolate the secondary voltage from the primary voltage.

(8) **Rain repellent**. The rain repellent is a kind of rain protection wind-screen coating developed can rapidly form a highly transparent and water repellent strong film on the wind-screen of automobiles or aircraft.

(9) **Thermistor**. A thermistor is a type of resistor whose resistance varies significantly with temperature, more so than in standard resistors. The word is a portmanteau of thermal and resistor. Thermistors are widely used as inrush current limiters, temperature sensors, self-resetting overcurrent protectors, and self-regulating heating elements.

(10) **Thermostat**. A temperature-sensitive electrical switch, which is used to control the temperature of the air inside the cabin. The thermostat contains an electrical switch that closes when the temperature drops to a preset value.

Current flowing through the closed switch opens the heater fuel valve, allowing the heater to begin heating the air. When the cabin air temperature rises to the upper value for which the thermostat is set, the switch opens, and the fuel valve shuts off the flow of fuel to the heater.

2. Answer the following questions in English.
(1) What parts of the aircraft need anti-icing?
(2) What means are used to prevent or control ice formation on aircraft?
(3) What are the three methods used to provide heated air?
(4) What is the function of the conductive coating in aircraft windows?
(5) What are the methods of windshield rain removal?

SUPPLEMENTARY READING

[1] Maintenance on pneumatic deicing systems varies with each aircraft model. The instructions of the airframe or system components manufacturer should be followed in all cases. Depending on the aircraft, maintenance usually consists of operational checks, adjustments, troubleshooting, and inspection.

[2] An operational check of the system can be made by operating the aircraft engines, or by using an external source of air. Most systems are designed with a test plug to permit ground checking the system without operating the engines. When using an external air source, make certain that the air pressure does not exceed the test pressure established for the system.

[3] Before turning the deicing system on, observe the vacuum operated instruments. If any of the gauges begin to operate, it is an indication that one or more check valves have failed to close and that reverse flow through the instruments is occurring. Correct the difficulty before continuing the test. If no movement of the instrument pointers occurs, turn on the deicing system.

[4] With the deicer system controls in their proper positions, check the suction and pressure gauges for proper indications. The pressure gauge will fluctuate as the deicer tubes inflate and deflate. A relatively steady reading should be maintained on the vacuum gauge. It should be noted that not all systems use a vacuum gauge. If the operating pressure and vacuum are satisfactory, observe the deicers for actuation.

[5] With an observer stationed outside the aircraft, check the inflation sequence to be certain that it agrees with the sequence indicated in the aircraft maintenance manual. Check the timing of the system through several complete cycles. If the cycle time varies more than is allowable, determine the difficulty and correct it. Inflation of the deicers must be rapid to provide efficient deicing. Deflation of the boot being observed should be completed before the next inflation cycle.

[6] Examples of adjustments that may be required include adjusting the deicing system control cable linkages, adjusting system pressure relief valves and deicing system vacuum (suction) relief valves.

[7] A pressure relief valve acts as a safety device to relieve excess pressure in the event of regulator valve failure. To adjust this valve, operate the aircraft engines and adjust a screw on the valve until the deicing pressure gauge indicates the specified pressure at which the valve should relieve. Vacuum relief valves are installed in a system that uses a vacuum pump to maintain constant suction during varying vacuum pump speeds. To adjust a vacuum relief valve, operate the engines. While watching the vacuum (suction) gauge, an assistant should adjust the suction relief valve adjusting screw to obtain the correct suction specified for the system.

[8] Not all troubles that occur in a deicer system can be corrected by adjusting system components. Some troubles must be corrected by repair or replacement of system components or by tightening loose connections. Note the probable causes and the remedy of each trouble listed in the chart. In addition to using troubleshooting charts, operational checks are sometimes necessary to determine the possible cause of trouble.

[9] During each preflight and scheduled inspection, check the deicer boots for cuts, tears, deterioration, punctures, and security; and during periodic inspections go a little further and check deicer components and lines for cracks. If weather cracking of rubber is noted, apply a coating of conductive cement. The cement in addition to sealing the boots against weather, dissipates static electricity so that it will not puncture the boots by arcing to the metal surfaces.

[10] The life of the deicers can be greatly extended by storing them when they are not needed and by observing these rules when they are in service:

(1) Do not drag gasoline hoses over the deicers.

(2) Keep deicers free of gasoline, oil, grease, dirt and other deteriorating substances.

(3) Do not lay tools on or lean maintenance equipment against the deicers.

(4) Promptly repair or resurface the deicers when abrasion or deterioration is noted.

(5) Wrap the deicer in paper or canvas when storing it.

[11] Thus far preventive maintenance has been discussed. The actual work on the deicers consists of cleaning, resurfacing, and repairing. Cleaning should ordinarily be done at the same time the aircraft is washed, using a mild soap and water solution. Grease and oil can be removed with a cleaning agent, such as naphtha, followed by soap and water scrubbing.

[12] Whenever the degree of wear is such that it indicates that the electrical conductivity of the deicer surface has been destroyed, it may be necessary to resurface the deicer. The resurfacing substance is a black, conductive neoprene cement. Prior to applying the resurfacing material, the deicer must be cleaned thoroughly and the surface roughened.

[13] Cold patch repairs can be made on a damaged deicer. The deicer must be relieved of its installed tension before applying the patch. The area to be patched must be clean and buffed to roughen the surface slightly.

[14] One disadvantage of a pneumatic deicer system is the disturbance of airflow over the wing and tail caused by the inflated tubes. This unwanted feature of the deicer boot system has led to the development of other methods of ice control, one of which is the thermal anti-icing system.

TOPIC 17：维修工作签字常用句

编号	英　文	中　文
1	Accomplished ××× task of EO according to work order ××× requirement. EO No. ××, Title：××	按×××号工作指令单的要求完成工程指令的×××项工作。工程指令编号：××,名称：××
2	2. After adjustment（replacement, reinstallation, reset circuit breaker）of ×××, Power-on-test (test run/inspection) OK	调节(更换、重新安装、重置跳开关)×××后,通电测试(试车/检查)正常
3	After confirmation from CHINA SOUTHERN (CHINA EASTERN/AIR CHINA) maintenance department, the problem has been solved according to the FAX dated ××, (is no influence to flight/comply with MEL) the affected aircraft is OK for further flight	请示南航(东航/国航)机务,该问题按××日期的传真进行处理(对飞行没有影响/符合MEL),该飞机可以继续飞行
4	After approval from CHINA SOUTHERN (CHINA EASTEN/AIR CHINA) maintenance department, the aircraft is cleared to dispatch	经南航(东航/国航)机务同意,放行飞机
5	Engine (APU) fail to start (start slow) was found by ground check	地面检查发现发动机(APU)起动不起来(起动慢)
6	Engineering task card ××（component monitoring sheet ××）accomplished	工程任务卡××(附件监控单××)完成
7	EO ×× accomplished. Title：××. P/N ××, S/N ×× checked	工程指令××已完成。名称：××,已检件号：××,序号××
8	EO (SB) No：×× accomplished. Title：××, check/test OK.	已完成工程指令(服务通告)××,标题：××,检查/测试正常
9	Replenished hydraulic fluid（nitrogen/oxygen）to standard for ××× on preflight (transit/postflight)	航前(过站/航后)对×××进行加液压油(充氮气/充氧气)至标准值
10	Inspection OK after tightening bolt（nut）	拧紧螺栓(螺帽)后,检查正常
11	It can be used after temporary treatment, and the failure will be thoroughly solved on postflight （while AOG）	做临时处理可以使用,待航后(停场时)彻底排除该故障
12	Nose wheel/main wheel/brake hub stab（worn）to limit was found by ground check	地面检查发现前轮/主轮/刹车毂刺伤/磨损超标
13	Recovered ×× which swapped to B-××××, PN：××, SN：××	将串至B—××××飞机上的××恢复,件号：××,序号：××
14	Since no spare part available, it will be replaced upon spare arrival	没有航材件,待来件后更换
15	Since the related spare part arrived, ×× failure dated on ×× was solved by ×× replacement. Power-on-test/test run/inspection per Maintenance Manual OK, and the related deferred failure item is cancelled	因相关航材到货,×月×日反映的××故障,更换××后,按维护手册要求,通电/试车/试验/检查正常,撤销相关故障保留

续表

编号	英　文	中　文
16	Since there is no spare part/no enough troubleshooting time/no applicable facility, and the failure complied with MEL (CDL)××-×× , deferred failure item was applied (renewed) for ×× days (hours, flights). Flight crew should perform the operation procedure according to related MEL (CDL) requirement	因没有航材件(排故时间不足、无可用设备)，且该故障符合MEL(CDL)××—××，已办理(续办)故障保留××天(小时、航班)，请机组按MEL(CDL)相关要求执行操作程序
17	The non-scheduled job card ×× accomplished on postflight/scheduled check. Title：××. Inspection OK per related requirement	航后/定检完成非例行工作卡××，标题：××。按相关要求检查正常
18	For fault isolation, ×× was swapped to B-×× ××, PN：××, SN：××	为故障隔离，将××串至B—××××飞机，件号：××，序号：××
19	For fault isolation, ×× was swapped from opposite side (from ×× aircraft/from Left/Right engine) on ground. Power-on-test/test-run/inspection per Maintenance Manual OK	为故障隔离，地面本机对调××(与××飞机对调、左/右发动机对调)，按维护手册要求，通电(试车)试验(检查)正常
20	Upon check, fault was found in ×××. After replacing ×××, power-on-test/test-run/inspection per Maintenance Manual OK. P/N ××× brand new (rotable); S/N: installed: ×××, removed: ×××; ATA: ××-××; Worker: ××; Inspector: ××; Station: ××; Date: ××	经检查×××存在故障，更换×××后，按维护手册要求，通电测试/试车/检查正常。件号：×××全新(周转件)；序号：装上：×××，拆下：×××；ATA：××—××；工作者：××；检查者(检验者)：××；地点：××；日期：××
21	×× failure (fuel/oil leakage, drained fuel/oil exceed limit, no bleed pressure, oil pressure low, fuel pressure high, no hydraulic indication, tire pressure indication swing, inoperative, low signal, fail to open, fail to retract, fail to extend, jam, damage, water leakage, fault code appears, circuit breaker pop-up, △ value excess limit) was found by ground check	地面检查发现××故障(漏油、余油超标、无引气压力、滑油压力低、燃油压力高、液压无指示、轮胎压力指示摆动、不工作、信号弱、打不开、收不上、放不下、卡阻、损坏、漏水、有故障代码、跳开关跳出、差值超标)
22	×× corrosion is found by visual inspection	目视检查发现××腐蚀
23	×× test failed on ground check, (×× test displays the message：×× clearance exceed limit upon ground measuring)	地面进行××测试时不通过(进行××测试显示信息：地面测量××间隙超标)
24	×× worn to limit was found by ground check (×× clearance exceed limit upon ground measuring)	地面检查发现××磨损超标(地面测量××间隙超标)
25	××× replaced without test	更换×××，测试未做

UNIT 18

INERTING GAS SYSTEM

[1] Three elements are required to initiate and sustain combustion: an ignition source (heat), fuel and oxygen. Combustion may be prevented by reducing any one of these three elements. If the presence of an ignition source can not be prevented within a fuel tank, then the tank may be made inert by:
- reducing the oxygen concentration of the ullage (the space above a liquid fuel) to below that capable of combustion (the combustion threshold);
- reducing the fuel concentration of the ullage to below the "lower explosive limit" (LEL), the minimum concentration capable combustion;
- increasing the fuel concentration to above the "upper explosive limit" (UEL), the maximum concentration capable of combustion.

[2] The risk of fuel tank explosions has been a common concern of aircraft manufacturers and airlines since the beginning of commercial flight. There are three major causes of fuel tank explosions as noted by FAA: the first source is electrical arcing that may occur if a part of the insulation on the electrical wire bundles that run through the fuel tanks wears off, exposing a small patch of bare wire, often not visually discernable, that contacts a metal surface; the second are friction sparks that can occur from rotating components, such as a steel fuel pump impeller, rubbing on the pump inlet check valve; the final source is auto ignition, where combustion can occur when the fuel/air mixture spontaneously ignites from heat; even in the absence of an ignition source.

[3] At present, inflammable vapors in fuel tanks are rendered inert by replacing the air in the tank with an inert gas, such as nitrogen, nitrogen enriched air, steam or carbon dioxide. This reduces the oxygen concentration of the ullage to below the combustion threshold. Alternate methods based on reducing the ullage fuel-air ratio to below the LEL or increasing the fuel-air ratio to above the UEL have also been proposed.

[4] An inert gas is a gas which does not undergo chemical reactions under a set of given conditions. The noble gases and nitrogen often do not react with many substances. Inert gases are used generally to avoid unwanted chemical reactions degrading a sample. These undesirable chemical reactions are often oxidation and hydrolysis reactions with

the oxygen and moisture in air. The term inert gas is context-dependent because nitrogen gas and several of the noble gases can be made to react under certain conditions.

[5] Purified nitrogen and argon gases are most commonly used as inert gases due to their high natural abundance (78% N_2, 1% Ar in air) and low relative cost. Unlike noble gases, an inert gas is not necessarily elemental and is often a compound gas. Like the noble gases the tendency for non-reactivity is due to the valence, the outermost electron shell, being complete in all the inert gases. This is a tendency, not a rule, as noble gases and other "inert" gases can react to form compounds.

[6] An airplane fuel tank inerting system provides an inert atmosphere in a fuel tank to minimize explosive ignition of fuel vapor. This AIR deals with the three methods of fuel tank inerting systems currently used in operational aircraft: ①on-board inert gas generation systems (OBIGGS); ② liquid/gaseous nitrogen systems; ③ Halon systems. The OBIGGS and nitrogen systems generally are designed to provide full-time fuel tank fire protection; the Halon systems generally are designed to provide only on-demand or combat-specific protection. This AIR does not treat the subject of explosion suppression foam (ESF) that has been used for fuel tank explosion protection on a number of military aircraft. ESF is a totally passive, full-time protection system. The primary disadvantages of foam are weight, reduction of usable fuel, and the added maintenance complexity when the foam must be removed for tank maintenance or inspection.

[7] The OBIGGS process starts with processed air, which is conditioned to optimum pressure and temperature and then passed through a series of air separation modules (ASM). The ASMs separate the oxygen from the process air through a permeable gas separation membrane, producing primarily nitrogen-enriched air (NEA). The NEA is delivered to the fuel tank where it displaces the flammable fuel/air mixture to reduce the risk of an explosion or fire.

[8] The system consists of either an independent thermal control unit (TCU) or one that is integrated with the environmental control system (ECS), along with the air separation unit (ASU) and instrumentation to control the operation and distribution of air to the fuel tanks.

[9] The following is a typical list of components that are used in the fuel tank inerting system:
- Shut-off valve;
- Heat exchanger to condition the temperature;
- High efficiency ozone converter to ensure long ASM life;
- Turbo compressor to condition the pressure (not required for all applications);
- Valves used to control NEA distribution and flow modes;
- Pressure, temperature and oxygen sensors;
- Contaminant filter;
- Air separation modules;

- Honeywell digital controller with full built-in-test display.

[10] On B737NG aircraft, the nitrogen generation system (NGS) is used, see Fig. 18-1. The nitrogen generation system uses bleed air from the left side of the pneumatic manifold. The NGS shutoff valve controls the airflow from the manifold. The NGS controller uses system pressure and voltage to control the NGS shutoff valve. The system does these functions:

- Controls the air pressure into the system;
- Changes the ozone in the air to oxygen;
- Decreases the temperature of the air;
- Removes contamination from the air;
- Removes oxygen from the air;
- Supplies nitrogen enriched air to the center tank;
- Does a check of system performance.

[11] The Nitrogen Generation System has these subsystems:

- Thermal control unit (TCU);
- Nitrogen generation;
- Distribution;
- Control;
- Indication.

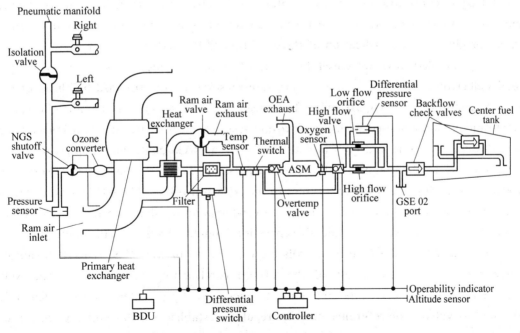

Fig. 18-1 Nitrogen generation system

[12] **The nitrogen generation system** gets bleed air from the left side of the pneumatic manifold. A sensor on the bleed air duct sends pressure data to the NGS controller. The

controller adjusts the NGS shutoff valve. The NGS shutoff valve controls the pressure that comes into the system. Bleed air goes through the catalytic converter to change the ozone in the air to oxygen. Ozone can decrease the performance and mechanical properties of the air separation module.

[13] **The heat exchanger** uses ram air to decrease the bleed air temperature to (160 ± 10)°F. The ram air valve adjusts the quantity of cool air that goes through the heat exchanger. The temperature sensor sends temperature data to the controller. The controller adjusts the ram air valve. The filter removes contamination before the air goes into the air separation module. A differential pressure sensor monitors the filter.

[14] **The air separation module** decreases the oxygen in the air below the quantity necessary to support combustion. The air separation module removes oxygen from the air and releases it overboard. Nitrogen enriched air (NEA) goes through the high flow valve to the center tank. The high flow valve controls the quantity of NEA that goes to the center tank. The controller uses data from the altitude sensor, differential pressure sensor, and airplane systems to open or close the valve.

[15] **The NEA distribution system** (NEADS) sends NEA to the center wing tank. NEA goes into the tank through an ejector nozzle in the climb vent in the left part of the tank. A float valve in the right part of the tank makes sure that the concentration of NEA is constant. A cross vent check valve makes sure that ambient air does not dilute the nitrogen concentration in the center tank during the descent. The controller monitors and controls system operating temperatures and pressures. An operability indicator gives a visual indication of the condition of the system.

[16] **The bleed pressure sensor** monitors the bleed air inlet pressure by an airtight sealed electronic circuitry. The bleed pressure sensor is on the forward bulkhead of the left air conditioning compartment. The bleed pressure sensor has an airtight sealed electronic circuitry housed within a corrosion-resistant steel shell. Bleed air comes from the pressure sense line and goes into the sensor. The NGS controller supplies power to the sensor which then monitors the bleed air pressure. When the bleed air pressure goes above 67 psig (462 kPa), a signal is sent to the NGS controller. The NGS controller then commands the NGS shutoff valve and the over-temperature shutoff valve (OTSOV) to close. Damage is prevented to the air separation module (ASM) and the center tank.

[17] **The NGS shutoff valve** controls bleed air flow for the NGS. The NGS shutoff valve is in the forward section of the left air conditioning compartment. The NGS shutoff valve is an electrically commanded, pneumatically actuated pressure regulating and shutoff valve. The reference pressure regulator establishes a constant gage pressure. The butterfly closure element is spring-loaded closed. A double acting, dual area, diaphragm-piston actuator positions the butterfly by comparing reference pressure to the downstream bleed air pressure admitted through a downstream sense tube. The solenoid controls the shutoff valve. The NGS shutoff valve includes a manual override.

[18] **The ozone converter** removes ozone from the engine bleed air by a catalytic process that converts ozone to oxygen. This process protects the air separation module (ASM) from oxidation of the membrane materials by ozone. The ozone converter is in the forward section of the left air conditioning compartment. The ozone converter uses a stainless steel mantle shell. The catalytic reactor core features a straight-channel tin configuration to minimize pressure drop. Hot bleed air that contains ozone enters the ozone converter through the inlet. As the air comes in contact with the heated catalytic reactor core the ozone is converted to oxygen. The bleed air then exits the converter through the outlet.

[19] **Heat exchanger:** The primary function of the heat exchanger (Fig. 18-2) is to condition the bleed air to $(160.0 \pm 10.08)°F$ $((71.1 \pm 5.68)°C)$. Hot, compressed air, bled from the bleed air system, is cooled by the heat exchanger on its way to the air separation module (ASM). In addition, the heat exchanger is designed to prevent hot bleed air from entering the fuel tank in the event of a system double failure. The heat exchanger is in the left ram air duct compartment, inboard of the ASM. The heat exchanger is an aluminum plate-fin, single-pass, crossflow, air-to-air heat exchanger. Both bleed air and ram air fins are made from aluminum alloy sheet.

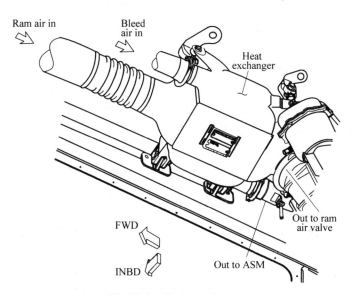

Fig. 18-2 Heater exchanger

NEW WORDS AND EXPRESSIONS

(1) inert[ɪ'nɜːt]　　　　　　　　　　惰性的,不活泼的
(2) element['elɪmənt]　　　　　　　要素;元件,零件;电池
(3) presence['prezəns]　　　　　　　存在,出现;出席,到场;仪表,仪态

(4) ullage['ʌlɪdʒ]　　　　　　　　　　　容器油面以上的空间
(5) threshold['θreʃhəʊld]　　　　　　　阈值,门槛值
(6) LEL：lower explosive limit　　　　爆炸下限
(7) UEL：upper explosive limit　　　　爆炸上限
(8) electrical arcing　　　　　　　　　电弧
(9) bundle['bʌndl]　　　　　　　　　捆,包,束
(10) patch[pætʃ]　　　　　　　　　　补片,补丁；斑,块；小块,小片
(11) bare[bɛə]　　　　　　　　　　　裸露的,赤裸的；
(12) discernable[dɪ'sə:nəbl]　　　　　　可辨别的,可认识的
(13) spontaneously[spɒn'teniəslɪ]　　　自发地,不由自主地
(14) render['rendə]　　　　　　　　　给予；呈递,提供
(15) enrich[ɪn'rɪtʃ]　　　　　　　　　使富有；使富裕；充实；使丰富,富集
(16) noble gas　　　　　　　　　　　惰性气体
(17) hydrolysis[haɪ'drɒlɪsɪs]　　　　　水解
(18) argon['ɑ:gɒn]　　　　　　　　　氩
(19) valence['veɪləns]　　　　　　　　(化合)价,原子价
(20) outermost['aʊtə,məʊst]　　　　　最外面的,离中心最远的
(21) electron[ɪ'lektrɒn]　　　　　　　电子
(22) OBIGGS：on-board inert gas generation systems　机载惰性气体发生系统
(23) ASM：air separation modules　　空气分离组件
(24) permeable['pə:mɪəbl]　　　　　　可渗透的,具渗透性的
(25) TCU：thermal control unit　　　　温度控制组件
(26) ECS：environmental control system　环境控制系统
(27) ASU：air separation unit　　　　　空气分离装置
(28) NGS：nitrogen generation system　氮气发生系统
(29) catalytic[,kætə'lɪtɪk]　　　　　　　接触反应的
(30) ozone['əʊzəʊn]　　　　　　　　　臭氧
(31) NEA：nitrogen enriched air　　　　富氮空气
(32) NEADS：NEA distribution system　NEA 分配系统
(33) ejector[ɪ'dʒektə]　　　　　　　　引射泵,喷射器
(34) airtight['ɛətaɪt]　　　　　　　　　气密的,密封的
(35) OTSOV：over-temperature shutoff valve　超温关断活门
(36) membrane['membreɪn]　　　　　　薄膜,隔膜
(37) mantle['mæntl]　　　　　　　　　覆盖物

EXERCISES

1. Translate the followings into Chinese.

(1) **Inert gas**. A gas that has a stable atomic structure. Its valence shell neither tries

to gain nor lose electrons. An inert gas has no tendency to unite with other elements to form chemical compounds. Helium, neon, argon, krypton, xenon, and radon are inert gases.

(2) **Ejector**. A form of jet pump used to pick up a liquid and move it to another location. Ejectors are used to ensure that the compartment in which the boost pumps are mounted is kept full of fuel. Part of the fuel from the boost pump flowing through the ejector produces a low pressure that pulls fuel from the main tank and forces it into the boost pump sump area.

(3) **Ignition (heat engine operation)**. The procedure by which the temperature of a portion of the fuel-air mixture in the engine is raised high enough for the hydrocarbon fuel to begin combining with the oxygen in the air. This heat is normally provided by an electrical spark.

(4) **FAA (federal aviation administration)**. The division within the Department of Transportation of the United States government that has the responsibility of promoting safety in the air, by both regulation and education. Aircraft and the airmen who operate them are licensed by the FAA, and the FAA maintains the airways along which the aircraft fly. The FAA is headed by a civilian, the Administrator of Federal Aviation.

(5) **Fuel-air mixture ratio**. The ratio of the number of pounds of fuel to the number of pounds of air in the mixture burned, in an internal combustion engine.

(6) **Access panel**. An easily removable panel that allows access to some portion of an aircraft structure for inspection and maintenance.

(7) **Access door**. A door which provides access to the inside of an aircraft structure.

(8) **Heat exchanger**. A device used to exchange heat from one medium to another. Radiators, condensers, and evaporators are all examples of heat exchangers. Heat always moves from the object or medium having the greatest level of heat energy to a medium or object having a lower level. Air-to-air, air-to-liquid, liquid-to-air, and liquid-to-liquid heat exchangers are commonly used with aircraft engines.

(9) **Shut-off valve (fluid power system component)**. A valve used in a fluid power system to shut off the flow of fluid.

(10) **Filter (fluid filter)**. A component in a system that traps and holds contaminants carried in the fluid flowing through the system. Aircraft fuel systems, lubrication systems, and hydraulic systems all use filters.

2. Answer the following questions in English.

(1) What are the ways to inert the fuel tank?

(2) What are the main causes of the fuel tank explosions?

(3) Which kind of gases can be used for inerting the fuel tank?

(4) Which elements are included in the nitrogen generation system?

(5) What is the function of the heat exchanger in NGS?

(6) Why is the NGS used to aircraft fuel tank venting system?

SUPPLEMENTARY READING

[1] Fuel tanks for combat aircraft have long been inerted, as well as self-sealing, but those for transport aircrafts, both military and civilian, have not, largely due to cost and weight considerations.

[2] Cleve Kimmel first pitched an inerting system to passenger airlines in the early 1960s. His proposed system for passenger aircraft would have used nitrogen. However, the Federal Aviation Administration refused to consider Kimmel's system after the airlines complained it was impractical. Indeed, early versions of Kimmel's system weighed 2,000 pounds, which would have probably made an aircraft too heavy to fly with passengers on it. However, the FAA did almost no research into making fuel tanks inert for 40 years, even in the face of several catastrophic fuel tank explosions. Instead, the FAA focused on keeping ignition sources out of the fuel tanks.

[3] The FAA did not consider lightweight inerting systems for commercial jets until the 1996 crash of TWA Flight 800. The crash was blamed on an explosion in the center wing fuel tank of the B747 used in the flight. This tank is normally used only on very long flights, and little fuel was present in the tank at the time of the explosion. A small amount of fuel in a tank is more critical than a large amount, since heat entering the fuel tank with residual fuel causes the fuel to increase in temperature faster and evaporate. This causes the ullage fuel air ratio to increase rapidly and the ullage fuel air ratio to exceed the lower flammability limit. Large quantity of fuel (high mass loading) in the fuel tank retains the heat energy and slows down the fuel evaporation rate. Explosion of Thai Airways International B737 in 2001 and Philippine Airlines B737 in 1990 also occurred in a tank that had residual fuel. All the above three explosions occurred on a warm day, in the Center Wing tank (CWT) that is within the contours of the fuselage. These fuel tanks are located in the vicinity of external equipment that heats the fuel tanks. The National Transportation Safety Board's (NTSB) final report on the crash of TWA B747 concluded "The fuel air vapor in the ullage of the TWA flight 800 CWT was flammable at the time of the accident." NTSB identified "Elimination of Explosive Mixture in Fuel Tanks in Transport Category Aircraft" as Number 1 item on its Most Wanted List in 1997.

[4] After the Flight 800 crashed, a 2001 report by an FAA committee stated that U.S. airlines would have to spend US $ 35 billions to retrofit their existing aircraft fleets with inerting systems that might prevent future such explosions. However, another FAA

group developed a nitrogen enriched air (NEA) based inerting system prototype that operated on compressed air supplied by the aircraft's propulsive engines. Also, the FAA determined that the fuel tank could be rendered inert by reducing the ullage oxygen concentration to 12% rather than previously accepted threshold of 9%-10%. Boeing commenced testing a derivative system of their own, performing successful test flights in 2003 with several B747 aircraft. The new, simplified inerting system was originally suggested to the FAA through public comment. It uses a hollow fiber membrane material that separates supplied air into nitrogen-enriched air (NEA) and oxygen enriched air (OEA). This technology is extensively used for generating oxygen-enriched air for medical purposes. It uses a membrane that preferentially allows the nitrogen molecule (molecular weight 28) to pass through it and not the oxygen molecule (molecular weight 32).

[5] Unlike the inerting systems on military aircraft, this inerting system would run continuously to reduce fuel vapor flammability whenever the aircraft's engines are running; and its goal is to reduce oxygen content within the fuel tank to 12%, lower than normal atmospheric oxygen content of 21%, but higher than that of inerted military aircraft fuel tanks, which is a target of 9% oxygen. This is accomplished by ventilating fuel vapor laden ullage gas out of the tank and into the atmosphere.

[6] Two other methods in current use to inert fuel tanks are a foam suppressant system and a ullage system. The FAA has decided that the added weight of both alternatives make them impractical for implementation in the aviation field. Some US military aircraft still use nitrogen based foam inerting systems, and some companies will ship containers of fuel with an ullage system across train routes.

TOPIC 18：工程指令编写规范

工程指令（engineering order，EO）：用于保证飞机/发动机/部附件持续适航性的指令性文件，主要针对飞机/发动机/部附件等，根据适航指令/服务通告/服务信函以及局方和运营人要求，用于执行改装、手册修订、修理和检查的指令。

> 对于符合 CAAC 适航指令适用性的飞机、发动机和部/附件，必须编制相应的工程指令予以执行。
> 对于决定执行的制造/协作厂家的服务通告、服务信函和电传等技术文件，必须以其为依据编制相应的工程指令予以执行。
> 对于决定结合送修执行的制造/协作厂家的服务通告、服务信函和电传等技术文件，必须以其为依据编制相应的执行通知单予以执行。
> 依据本单位维修工作的需要为提高机群的安全性、经济性、可靠性和功能性，依据有关技术文件，可编制颁发工程指令予以执行。
> 工程指令编号原则：对于 A320 机型，B（某航空公司机队代码）+ 机型（320 = A320）+ ATA 章号（两位）+ 发动机代码（仅为发动机章节的 EO 显示，P = PW4000，I = IAE，C = CFM）+ 顺序号（001，…）+ 版本号（R0，…），如 B32027001R0，B32072I028R2。
> 适用范围、施工方法、检查间隔、完成时限应与适航资料保持一致，如涉及到航空器适航性和飞行安全的服务通告/服务信函/有相关的适航指令，则适用范围、施工方法、检查间隔、完成时限必须符合适航指令的要求，且适航资料内包含的"警告""注意"等内容也应编入 EO 中，如采用替代方法或调整完成时间必须通过客户获得相应的适航当局批准。

表 T18-1　工程指令样例

EO 编号 NO (1)		中国××航空公司 工程指令 CHINA ×× AIRLINES ENGINEERING ORDER		编写 PREPARE (17)		
修订号 REV NO (2)				日期 DATE (18)		
机型 A/C TYPE (3)				审核 REVIEW (19)		
期限 DUE TATE (4)				日期 DATE (20)		
停场时间 GND (5)		CAD 号 (11)	REV	批准 APPROVE (21)		
预计工时 EST MHR (6)		AD 号 (12)	REV	日期 DATE (22)		
计划 SCHEDULE (7)		SB 号 (13)	REV	PAGE OF (23)		
MSL(8)	☐ YES ☐ NO	其他 OTHER(14)		PDN (24)	☐ YES	☐ NO
PCR(9)	☐ YES ☐ NO	重大改装 MAJOR ALTER (15)	☐ YES ☐ NO	飞机改装 A/C MOD (25)	☐ YES	☐ NO
RII(10)	☐ YES ☐ NO	重大修理 MAJOR REPAIR (16)	☐ YES ☐ NO	部件改装 PART MOD (26)	☐ YES	☐ NO
标题 SUBJECT (27)						
适用范围 EFFECTIVITY (28)						
受影响的部件 COMPONENTS AFFECTED (29)						
受影响的备件 SPARES AFFECTED (30)						
颁布本指令的原因 REASON FOR THE ENGINEERING ORDER (31)						
重量与平衡 WT&BAL (32)		重量 WT(LBS)_____		力矩 MT(LB-INS)		
备注 REMARKS (33)						

(1) 编号:EO 编号。

(2) 修改号:EO 的修改版代号,用 R1,R2,R3,…,Rn 表示。

(3) 机型:飞机型号,如 A320。

(4) 期限:完成本 EO 的最后期限,以飞行小时、起落数、日历时间或诸如"下次飞行前"等表示,但不能使用"尽快""立即"等类似的字眼。对于结合送修或大修执行的 EO,写上"结合送修或大修完成"。

(5) 停场时间:执行指令飞机所需的停场时间,必须以小时为单位,为通告给出时间的 2 倍。

(6) 工时:完成本指令所需工时,为通告给出工时的 2.5 倍。

(7) 计划:本指令的执行计划,如:结合 C 检、航后等。

(8) MSL:维护信函,按实际情况选择 YES 或 NO,在相应的方框内打×。

(9) PCR:技术资料修改,按实际情况选择 YES 或 NO,在相应的方框内打×。

(10) RII:必检要求项目。

(11) CAD 号:CAD 编号。

(12) AD 号:AD 的编号。

(13) SB 号:SB 或 ASB 的编号。

(14) 其他:本 EO 依据文件(SL、TELE 等)的编号。

(15) 重大改装:按实际情况选择 YES 或 NO,在相应的方框内打×。

(16) 重大修理:按实际情况选择 YES 或 NO,在相应的方框内打×。

(17) 编写:编写者签名。

(18) 日期:编写日期。

(19) 审核:审核者签名。

(20) 日期:审核日期。

(21) 批准:批准者签名。

(22) 日期:批准日期。

(23) 页码:填入总页数和本页。

(24) PDN:零部件处理通知单,按实际情况选择 YES 或 NO,在相应的方框内打×。

(25) 飞机改装:按实际情况选择 YES 或 NO,在相应的方框内打×。

(26) 部件改装:按实际情况选择 YES 或 NO,在相应的方框内打×(注:发动机的改装属于部件改装)。

(27) 标题:EO 的标题。

(28) 适用范围:针对飞机下发的 EO,填写飞机注册号;针对发动机下发的 EO,填写发动机型号和序号;针对部件下发的 EO,填写部件名称、件号和序号。需与 AD、SB 适用性保持一致,如适用性不一致需在 AD、SB 评估单上有所体现。

(29) 受影响部件:根据 EO 依据文件,填入需改装或返工的部件的名称、件号和序号。有些部件在 A320 和 A330 等其他空客机型上同时安装,在执行相关部件改装时需注意。

(30) 受影响的备件:填入目前库存的需改装或返工的备件名称、件号和序号,并写明目前存库地点。

(31) 颁发本指令的理由：简述颁发本指令的原因和背景。

(32) 重量与平衡：完成 EO 后重量和力矩的改变量,增加用(+),减少用(-)。

(33) 备注：用于额外的说明(包括生产方面、航材方面、技术方面。如改版还应包括修订原因、修订要点,与上版指令替代关系等。如涉及 AD,应说明 EO 所依据 AD 规定措施的具体段落,与受影响其他 EO 之间的替代关系等),按需要填写。

表 T18-2　工程指令样例(续 1)

EO 编号 NO (1)	中国××航空公司工程指令 CHINA ×× AIRLINES ENGINEERING ORDER	PAGE OF (23)
修订号 REV NO (2)		
标题 SUBJECT (27)		
计划资料 PLANNING MATERIAL		
计划安排 SCHEDULE(34)		
器材 MATERIAL(35)		
专用工具或设备 SPECIAL TOOLS AND EQUIPMENT(36)		
图纸 DRAWING(37)		
参考资料 REFERENCR (38)		

(34) 计划安排：执行本指令在计划安排上的特别要求或说明。

(35) 器材：单部飞机/发动机/部件执行本 EO 所需器材的名称、件号和数量。用于指导生产准备 EO 所需器材及工作者领取 EO 所需器材。包括工作所需改装包,消耗品等。

(36) 专用工具或设备：执行本 EO 所需的专用工具或设备的名称、件号和数量。包括检查、改装、测试程序中所需的专用工具和设备。如特殊力矩的磅表、高压的地面氮气源等。

(37) 图纸：执行本 EO 所需图纸的名称和图号。

(38) 参考资料：执行 EO 所需手册、文件等的名称和章节号。

表 T18-3　工程指令样例(续 2)

EO 编号 NO (1)	中国××航空公司工程指令 CHINA ×× AIRLINES ENGINEERING ORDER			PAGE OF (23)	
修订号 REV NO (2)					
标题 SUBJECT (27)					
航材资料 PLANNING MATERIAL					
EO 完成期限 DUE DATE(4)					
以下器材用于完成/THE FOLLOWING ARE USED TO ACCOMPLISH (39)					
☐ AD/CAD	☐ ASB	☐ SB	☐ SL/TEL	☐ OTHER	
客户所定器材 MATERIAL ORDERED BY CUSTOMERS 件号 P/N	名称 NOMENCLATURE	数量 QTY	单价 PRICE	厂家名称及代码 VENDOR NAME & CODE	
(40)	(40)	(40)	(40)	(40)	

(39) 器材的用途：根据实际情况,在相应的方框内打 ×。

(40) 客户所订器材：客户订购的用于改装的器材和设备的件号、名称、数量、单价、厂家名称及代码,用于航材订购。

表 T18-4　工程指令样例（续 3）

EO 编号 NO（1）	中国××航空公司工程指令 CHINA ×× AIRLINES ENGINEERING ORDER		PAGE OF（23）
修订号 REV NO（2）			
标题 SUBJECT（27）			
施工说明 ACCOMPLISHMEN INSTRUCTION		工作者 MECH	检验员 INSP
执行 EO 前的注意事项 CAUTIONS OF EO IMPLEMENTATION （41）		（43）	（44）
（42）			

（41）执行 EO 前的注意事项：工作者在执行 EO 前必须执行的步骤，已印刷好，无须填写。

（42）施工说明：执行 EO 的施工步骤。

（43）工作者：工作者在画线处签名。

（44）检验者：检验者在画线处签名。

UNIT 19

ENGINE CONSTRUCTION AND PRINCIPLE

[1] For an aircraft to remain in level flight, thrust must be provided that is equal to and in the opposite direction of the aircraft drag. This thrust, or propulsive force, is provided by a suitable type of aircraft heat engine. All heat engines have in common the ability to convert heat energy into mechanical energy by the flow of some fluid mass (generally air) through the engine. In all cases, the heat energy is released at a point in the cycle where the working pressure is high relative to atmospheric pressure. In a reciprocating engine, the functions of intake, compression, combustion, and exhaust all take place in the same combustion chamber. Consequently, each must have exclusive occupancy of the chamber during its respective part of the combustion cycle. A significant feature of the gas turbine engine is that separate sections are devoted to each function, and all functions are performed simultaneously without interruption.

[2] Turbine engines are classified according to the type of compressors they use. There are three types of compressors: centrifugal flow, axial flow, and centrifugal-axial flow. Compression of inlet air is achieved in a centrifugal flow engine by accelerating air outward perpendicular to the longitudinal axis of the machine. The axial-flow engine compresses air by a series of rotating and stationary airfoils moving the air parallel to the longitudinal axis. The centrifugal-axial flow design uses both kinds of compressors to achieve the desired compression. The path the air takes through the engine and how power is produced determine the type of engine. There are four types of aircraft turbine engines: turbojet, turboprop, turbofan, and turboshaft.

[3] **Turbojet**: The turbojet engine consists of four sections: compressor, combustion chamber, turbine section, and exhaust. The compressor section passes inlet air at a high rate of speed to the combustion chamber. The combustion chamber contains the fuel inlet and igniter for combustion. The expanding air drives a turbine, which is connected by a shaft to the compressor, sustaining engine operation. The accelerated exhaust gases from the engine provide thrust. This is a basic application of compressing air, igniting the fuel-air mixture, producing power to self-sustain the engine operation, and exhaust

for propulsion. Turbojet engines are limited in range and endurance. They are also slow to respond to throttle applications at slow compressor speeds.

[4] **Turboprop**: A turboprop engine is a turbine engine that drives a propeller through a reduction gear. The exhaust gases drive a power turbine connected by a shaft that drives the reduction gear assembly. Reduction gearing is necessary in turboprop engines because optimum propeller performance is achieved at much slower speeds than the engine's operating rpm. Turboprop engines are a compromise between turbojet engines and reciprocating powerplants. Turboprop engines are most efficient at speeds between 250 and 400 mph and altitudes between 18,000 and 30,000 feet. They also perform well at the slow airspeeds required for takeoff and landing, and are fuel efficient. The minimum specific fuel consumption of the turboprop engine is normally available in the altitude range of 25,000 feet to the tropopause.

[5] **Turboshaft**: It delivers power to a shaft that drives something other than a propeller. The biggest difference between a turbojet and turboshaft engine is that on a turboshaft engine, most of the energy produced by the expanding gases is used to drive a turbine rather than produce thrust. Many helicopters use a turboshaft gas turbine engine. In addition, turboshaft engines are widely used as auxiliary power units on large aircraft.

[6] **Turbofan**: Turbofans were developed to combine some of the best features of the turbojet and the turboprop. Turbofan engines are designed to create additional thrust by diverting a secondary airflow around the combustion chamber. The turbofan bypass air generates increased thrust, cools the engine, and aids in exhaust noise suppression. This provides turbojet-type cruise speed and lower fuel consumption. The inlet air that passes through a turbofan engine is usually divided into two separate streams of air(Fig.19-1). One stream passes through the engine core, while a second stream bypasses the engine core. It is this bypass stream of air that is responsible for the term "bypass engine". A turbofan's bypass ratio refers to the ratio of the mass airflow that passes through the fan divided by the mass airflow that passes through the engine core.

Fig. 19-1　Turbofan engine

[7] A typical turbine engine consists of an air inlet, compressor, combustion chambers, a turbine section, exhaust section, accessory section, and the systems necessary for starting, lubrication, fuel supply, and auxiliary purposes, such as anti-icing, cooling, and pressurization.

[8] **Air entrance**: The air entrance is designed to conduct incoming air to the compressor with a minimum energy loss resulting from drag or ram pressure loss; that is, the flow of air into the compressor should be free of turbulence to achieve maximum operating efficiency. Proper inlet design contributes materially to aircraft performance by increasing the ratio of compressor discharge pressure to duct inlet pressure. Since on most modern turbofan engines the huge fan is the first part of the aircraft the incoming air comes into contact with, icing protection must be provided. This prevents chunks of ice from forming on the leading edge of the inlet, breaking loose, and damaging the fan. Warm air is bled from the engine's compressor and is ducted through the inlet to prevent ice from forming. If inlet guide vanes are used to straighten the air flow, then they also have anti-icing air flowing through them. The inlet also contains some sound-reducing materials that absorb the fan noise and make the engine quieter.

[9] **Compressor section**: Its primary function is to supply air in sufficient quantity to satisfy the requirements of the combustion burners. To fulfill its purpose, the compressor must increase the pressure of the mass of air received from the air inlet duct, and then discharge it to the burners in the quantity and at the pressures required. A secondary function of the compressor is to supply bleed air for various purposes in the engine and aircraft. The exact location of the bleed ports is, of course, dependent on the pressure or temperature required for a particular job. Air is often bled from the final or highest pressure stage since, at this point, pressure and air temperature are at a maximum. The two principal types of compressors currently being used in gas turbine aircraft engines are centrifugal flow and axial flow.

[10] The centrifugal-flow compressor achieves its purpose by picking up the entering air and accelerating it outwardly by centrifugal action, it consists of an impeller (rotor), a diffuser (stator), and a compressor manifold. The impeller, whose function is to pick up and accelerate the air outwardly to the diffuser, may be either of two types: single entry or double entry. The principal differences between the two types of impellers are size and ducting arrangement. The diffuser is an annular chamber provided with a number of vanes forming a series of divergent passages into the manifold. The diffuser vanes direct the flow of air from the impeller to the manifold at an angle designed to retain the maximum amount of energy imparted by the impeller. They also deliver the air to the manifold at a velocity and pressure satisfactory for use in the combustion chambers. The compressor manifold diverts the flow of air from the diffuser, which is an integral part of the manifold, into the combustion chambers. The manifold has one outlet port for each chamber so that the air is evenly divided. these

outlet ducts perform a very important part of the diffusion process; that is, they change the radial direction of the airflow to an axial direction, in which the diffusion process is completed after the turn.

[11] The axial-flow compressor compresses air while the air continues in its original direction of flow, thus avoiding the energy loss caused by turns, it has two main elements: a rotor and a stator. The rotor, turning at high speed, takes in air at the compressor inlet and impels it through a series of stages. From inlet to exit, the air flows along an axial path and is compressed at a ratio of approximately 1.25 : 1 per stage. The action of the rotor increases the compression of the air at each stage and accelerates it rearward through several stages. With this increased velocity, energy is transferred from the compressor to the air in the form of velocity energy. The function of the stator vanes is to receive air from the air inlet duct or from each preceding stage and increase the pressure of the air and deliver it to the next stage at the correct velocity and pressure. They also control the direction of air to each rotor stage to obtain the maximum possible compressor blade efficiency.

[12] **Combustion section**: The combustion section houses the combustion process, which raises the temperature of the air passing through the engine. This process releases energy contained in the air/fuel mixture. The major part of this energy is required at the turbine or turbine stages to drive the compressor. The primary function of the combustion section is, of course, to burn the fuel/air mixture, thereby adding heat energy to the air. To do this efficiently, the combustion chamber must: ①Provide the means for proper mixing of the fuel and air to assure good combustion; ② Burn this mixture efficiently; ③ Cool the hot combustion products to a temperature that the turbine inlet guide vanes/blades can withstand under operating conditions; ④Deliver the hot gases to the turbine section.

[13] The location of the combustion section is directly between the compressor and the turbine sections. The combustion chambers are always arranged coaxially with the compressor and turbine regardless of type, since the chambers must be in a through-flow position to function efficiently. All combustion chambers contain the same basic elements: ①Casing; ②Perforated inner liner; ③Fuel injection system; ④Some means for initial ignition; ⑤ Fuel drainage system to drain off unburned fuel after engine shutdown.

[14] There are currently three basic types of combustion chambers, they are: ①can type; ②can-annular type; ③annular type. The can-type combustion chamber is typical of the type used on turboshaft and APUs, it consists of an outer case or housing, within which there is a perforated stainless steel (highly heat resistant) combustion chamber liner or inner liner. The can-annular combustion chamber is not used in modern engines. A number of flame tubes are fitted inside a common air casing. The annular combustion chamber consists of a single flame tube, completely annular in form, which is contained

in an inner and outer casing. The main advantage of the annular combustion chamber is that for the same power output, the length of the chamber is only 75 per cent of that of a can-annular system of the same diameter, resulting in a considerable saving in weight and cost. Another advantage is the elimination of combustion propagation problems from chamber to chamber.

[15] **Turbine section:** The turbine has the task of providing power to drive the compressor and accessories. It does this by extracting energy from the hot gases released from the combustion system and expanding them to a lower pressure and temperature. The continuous flow of gas to which the turbine is exposed may enter the turbine at a temperature between 850℃ and 1,700℃ which is far above the melting point of current materials technology.

[16] The turbine assembly consists of two basic elements: turbine inlet guide vanes and turbine disk. The stator element is known by a variety of names, of which turbine inlet nozzle vanes, turbine inlet guide vanes, and nozzle diaphragm are three of the most commonly used. The turbine inlet nozzle vanes are located directly aft of the combustion chambers and immediately forward of the turbine wheel. This is the highest or hottest temperature that comes in contact with metal components in the engine. The turbine inlet temperature must be controlled, or damage will occur to the turbine inlet vanes. When the turbine blades are installed, the disk then becomes the turbine wheel. There are various ways of attaching turbine blades, some similar to compressor blade attachment. The most satisfactory method utilizes the fir-tree design. The disk acts as an anchoring component for the turbine blades. Since the disk is bolted or welded to the shaft, the blades can transmit to the rotor shaft the energy they extract from the exhaust gases.

[17] A turbine stage consists of a row of stationary vanes or nozzles, followed by a row of rotating blades. In some models of turboprop engine, as many as five turbine stages have been utilized successfully. It should be remembered that, regardless of the number of wheels necessary for driving engine components, there is always a turbine nozzle preceding each wheel.

[18] **Exhaust Section:** Gas turbine engines for aircraft have an exhaust system which passes the turbine discharge gases to atmosphere at a velocity in the required direction, to provide the necessary thrust. The exhaust section is located directly behind the turbine section and ends when the gases are ejected at the rear in the form of a high-velocity exhaust gases. The design of the exhaust system, therefore, exerts a considerable influence on the performance of the engine. The cross sectional areas of the jet pipe and propelling or outlet nozzle affect turbine entry temperature, the mass flow rate, and the velocity and pressure of the exhaust jet.

[19] There are two types of exhaust nozzle designs: the converging design for subsonic gas velocities and the converging diverging design for supersonic gas velocities.

The exhaust nozzle opening may be of either fixed or variable area. The fixed-area type is the simpler of the two exhaust nozzles since there are no moving parts. The outlet area of the fixed exhaust nozzle is very critical to engine performance. If the nozzle area is too large, thrust is wasted; if the area is too small, the engine could choke or stall. A variable-area exhaust nozzle is used when an augmenter or afterburner is used due to the increased mass of flow when the afterburner is activated. It must increase its open area when the afterburner is selected. When the afterburner is off, the exhaust nozzle closes to a smaller area of opening.

[20] **Accessory Section**: The accessory section of the turbojet engine has various functions. The primary function is to provide space for the mounting of accessories necessary for operation and control of the engine. Generally, it also includes accessories concerned with the aircraft such as electric generators and fluid power pumps. Secondary functions include acting as an oil reservoir and/or oil sump, and housing the accessory drive gears and reduction gears.

[21] The arrangement and driving of accessories have always been major problems on gas turbine engines. Driven accessories are usually mounted on common pads either ahead of or adjacent to the compressor section, depending on whether the engine is centrifugal flow or axial flow. The components of the accessory section of all centrifugal flow and axial flow engines have essentially the same purpose even though they often differ quite extensively in construction details and nomenclature.

NEW WORDS AND EXPRESSIONS

(1) intake['ɪnteɪk]　　　　　　　　　进口,进气道
(2) centrifugal[ˌsen'trɪfjʊgl]　　　　离心的,离心机
(3) axial-flow　　　　　　　　　　　轴流
(4) exhaust[ɪg'zɔːst]　　　　　　　　排气
(5) range[reɪndʒ]　　　　　　　　　航程,射程,距离,范围
(6) endurance[ɪn'djʊərəns]　　　　　续航时间
(7) throttle['θrɑːtl]　　　　　　　　节气门,油门
(8) reduction gear　　　　　　　　　减速齿轮
(9) rpm: revolutions per minute　　每分钟转数
(10) mph: miles per hour　　　　　英里每小时
(11) specific fuel consumption　　　燃油消耗率
(12) tropopause['trɑpə,pɔz]　　　　对流层顶
(13) suppression[sə'preʃn]　　　　　抑制
(14) bypass ratio　　　　　　　　　涵道比
(15) lubrication[lʊbrɪkeɪʃən]　　　　润滑

(16) turbulence[tə:bjʊləns]　　紊流
(17) chunk[tʃʌŋk]　　相当大的量，大块，厚片
(18) guide vane　　导向叶片
(19) diffuser[dɪ'fju:zə]　　扩散器，扩压器
(20) divergent[daɪ'və:dʒənt]　　扩张的，发散的
(21) combustion[kəm'bʌstʃən]　　燃烧
(22) can type　　管型
(23) can-annular　　环管的
(24) annular['ænjələ]　　环形的
(25) fir-tree　　枞树型
(26) afterburner['æftəbə:nə]　　加力燃烧室

EXERCISE

1. Translate the followings into Chinese.

(1) **Endurance**. The length of time an aircraft can remain in the air. The power produced by the engines and the flight conditions can be regulated to give the aircraft the greatest speed, the greatest range, or the greatest endurance.

(2) **Throttle (aircraft engine control)**. The control in an aircraft that regulates the power or thrust the pilot wants the engine to develop.

(3) **Reduction gear train**. A gear arrangement in which the output shaft turns more slowly than the input shaft. A reduction gear train is used to increase the torque produced by a motor.

(4) **Specific fuel consumption**. A measure of the amount of fuel burned in a heat engine to produce a given amount of power. Specific fuel consumption of a reciprocating engine is expressed in pounds of fuel burned per hour for each horsepower produced.

(5) **Bypass ratio**. The ratio of the mass airflow, in pounds per second, through the fan section of a turbofan engine to the mass airflow that passes through the gas generator portion of the engine.

(6) **Turbulence**. A condition of fluid flow in which the flow is not smooth. The velocity of the flow changes rapidly, and the flow direction often reverses itself.

(7) **Axial flow compressor**. A type of compressor used in gas turbine engines in which the air passes though the compressor in essentially a straight line, parallel to the axis of the compressor. The compressor is made of a number of stages of rotating compressor blades between stages of stationary stator vanes. The compression ration is determined by the number of stages of compression.

(8) **Diffuser** (**gas turbine engine component**). A duct installed in the compressor outlet in a turbine engine to reduce the velocity of the air leaving the compressor and increase its pressure.

(9) **Fir-tree attachment**. A method of attaching turbine blades into the turbine wheel. The root of the blade is shaped like a notched-edge Christmas tree which fits loosely in a similar-shaped slot in the periphery of the wheel. When the engine is cold, the blades are loose, but as the engine heats up, the blades tighten in the wheel.

(10) **Afterburner** (**gas turbine engine component**). A device in the exhaust system of a turbojet or turbofan engine, used to increase the thrust for takeoff and for special flight conditions. It called reheaters in the United Kingdom, use a large amount of fuel, but the extra thrust they produce makes them efficient for high-performance aircraft.

2. **Answer the following questions in English.**
(1) What are the four types of turbine engine?
(2) What is the bypass ratio?
(3) What are the functions of combustion chamber?
(4) What are the basic elements of combustion chamber?
(5) What are the types of exhaust nozzle designs?
(6) What are the functions of accessory section of the gas turbine engine?

SUPPLEMENTARY READING

[1] **Turbine engine operating principles:** The principle used by a gas turbine engine as it provides force to move an airplane is based on Newton's Third Law. This law states that for every action there is an equal and opposite reaction; therefore, if the engine accelerates a mass of air (action), it applies a force on the aircraft (reaction). The turbofan generates thrust by giving a relatively slower acceleration to a large quantity of air. The old pure turbojet engine achieves thrust by imparting greater acceleration to a smaller quantity of air. This was its main problem with fuel consumption and noise.

[2] The mass of air is accelerated within the engine by the use of a continuous-flow cycle. Ambient air enters the inlet diffuser where it is subjected to changes in temperature, pressure, and velocity due to ram effect. The compressor then increases pressure and temperature of the air mechanically. The air continues at constant pressure to the burner section where its temperature is increased by combustion of fuel. The energy is taken from the hot gas by expanding through a turbine which drives the compressor, and by expanding through an exhaust nozzle designed to discharge the exhaust gas at high velocity to produce thrust.

[3] The high velocity gases from the engine may be considered continuous, imparting this force against the aircraft in which it is installed, thereby producing thrust. The formula for thrust can be derived from Newton's second law, which states that force is proportional to the product of mass and acceleration.

[4] Thrust of a gas turbine engine can be increased by two methods: increasing the mass flow of air through the engine or increasing the gas velocity. If the velocity of the turbojet engine remains constant with respect to the aircraft, the thrust decreases if the speed of the aircraft is increased. This is because V_1 increases in value. This does not present a serious problem, however, because as the aircraft speed increases, more air enters the engine, and jet velocity increases. The resultant net thrust is almost constant with increased airspeed.

[5] The Brayton cycle is the name given to the thermodynamic cycle of a gas turbine engine to produce thrust. This is a variable volume constant-pressure cycle of events and is commonly called the constant-pressure cycle. A more recent term is "continuous combustion cycle". The four continuous and constant events are intake, compression, expansion (includes power), and exhaust. These cycles are discussed as they apply to a gas-turbine engine. In the intake cycle, air enters at ambient pressure and a constant volume. It leaves the intake at an increased pressure and a decrease in volume. At the compressor section, air is received from the intake at an increased pressure, slightly above ambient, and a slight decrease in volume. Air enters the compressor where it is compressed. It leaves the compressor with a large increase in pressure and decrease in volume, created by the mechanical action of the compressor. The next step, expansion, takes place in the combustion chamber by burning fuel, which expands the air by heating it. The pressure remains relatively constant, but a marked increase in volume takes place. The expanding gases move rearward through the turbine assembly and are converted from velocity energy to mechanical energy by the turbine. The exhaust section, which is a convergent duct, converts the expanding volume and decreasing pressure of the gases to a final high velocity. The force created inside the engine to keep this cycle continuous has an equal and opposite reaction (thrust) to move the aircraft forward.

[6] Bernoulli's principle (whenever a stream of any fluid has its velocity increased at a given point, the pressure of the stream at that point is less than the rest of the stream) is applied to gas turbine engines through the design of convergent and divergent air ducts. The convergent duct increases velocity and decreases pressure. The divergent duct decreases velocity and increases pressure. The convergent principle is usually used for the exhaust nozzle. The divergent principle is used in the compressor and diffuser where the air is slowing and pressurizing.

[7] **Gas Turbine Engine Performance**: Thermal efficiency is a prime factor in gas turbine performance. It is the ratio of net work produced by the engine to the chemical

energy supplied in the form of fuel. The three most important factors affecting the thermal efficiency are turbine inlet temperature, compression ratio, and the component efficiencies of the compressor and turbine. Other factors that affect thermal efficiency are compressor inlet temperature and combustion efficiency. In actual operation, the turbine engine exhaust temperature varies directly with turbine inlet temperature at a constant compression ratio.

[8] Rpm is a direct measure of compression ratio; therefore, at constant rpm, maximum thermal efficiency can be obtained by maintaining the highest possible exhaust temperature. Since engine life is greatly reduced at high turbine inlet temperatures, the operator should not exceed the exhaust temperatures specified for continuous operation. In the previous discussion, it was assumed that the state of the air at the inlet to the compressor remains constant. Since this is a practical application of a turbine engine, it becomes necessary to analyze the effect of varying inlet conditions on the thrust or power produced. The three principal variables that affect inlet conditions are the speed of the aircraft, the altitude of the aircraft, and the ambient temperature. To make the analysis simpler, the combination of these three variables can be represented by a single variable called stagnation density.

[9] The power produced by a turbine engine is proportional to the stagnation density at the inlet. In addition, the thrust output increases since the air at reduced temperature has an increased density. The increase in density causes the mass flow through the engine to increase. The altitude effect on thrust can also be discussed as a density and temperature effect. In this case, an increase in altitude causes a decrease in pressure and temperature. Since the temperature lapse rate is lower than the pressure lapse rate as altitude is increased, the density is decreased. Although the decreased temperature increases thrust, the effect of decreased density more than offsets the effect of the colder temperature. The net result of increased altitude is a reduction in the thrust output.

TOPIC 19：科技论文写作规范

科技论文是科技人员介绍有关科研成果的文章，从结构上通常分为标题（title）、作者（author）、摘要（abstract）、关键词（key words）、正文部分（通常由引言（introduction）、主体（body）和结论（conclusion）组成）、致谢（acknowledges）以及参考文献（reference）7 个部分。

1. 论文标题

科技论文的标题应该能包含文章的主题与大意，以便于索引。一般情况下标题应力求简明扼要，没有主谓宾的固定结构，长度一般不超过 20 个词，句式应避免问句格式，且不使用不标准的缩略语。

2. 作者及工作单位

作者及工作单位放在标题的下方，各作者间以逗号隔开，不同单位的作者用上标表示出来。

3. 摘要

摘要是全文的浓缩和凝练。通过阅读摘要，读者能很快了解全文的概括，以决定是否阅读全文。摘要的长度，一般在 100 个单词左右或稍长一些，要保证重点突出、言简意赅、内容完整、结构严谨，避免过分简单和使用不标准的缩略语。此外，摘要中应不使用问句或感叹句。

摘要的内容包括：研究本课题的前提、目的、任务、范围与重要性，研究对象的特征；研究内容与所用原理、理论、手段和方法；主要结果与成果的意义、实用价值和应用范围；一般结论。

摘要不用图、表、化学结构式、非公知公用的符号和术语，只用标准科学术语的命名，应用第三人称。

英文摘要一般应用被动语态，字数一般不少于 10 个单句，约 500 单词。过去的研究工作用过去时，所得的结论用一般现在时。摘要中常用的句型和例句有：

(1) This paper treats (introduces) an important problem in...

(2) A new technique (method, manner) on... is showing this paper.

(3) This paper provides (shows, develops extends, describes) a new approach for...

(4) The author describes the technique of...

(5) It is suggested that some basic steps be taken in order to...

(6) This paper presents a thorough study of the output stability of pump.

(7) The purpose of this article is to explore the relation between sampling period and stability.

4. 关键词

关键词一般 3~6 个为宜，不能用句子或长句子成分来标引，不能把同义词或近义词并列为关键词。关键词具有以下 4 个特点：

(1) 代表性：最能反映论文的中心内容，表征其性质；
(2) 专指性：词意单一，指向性强，特异性高；
(3) 可检索性：最具有实质意义的检索语言；
(4) 规范性：选用规范的检索语言。

5. 引言（前言）

文章的开头及引言部分，主要介绍与文章有关的背景知识以及现存的问题，从而引出论文的写作目的和主题，内容一般包括：

(1) 背景与目的：前人研究的进展及评价，现在的知识空白和没有解决的问题，本课题的缘起、性质、重要性，本研究的理由、目的、作用、意义及在相关领域的地位。

(2) 任务与范围：课题所要解决的问题和研究任务，说明采用的研究方法和途径，理论依据和实验基础以及研究所涉及的范围、规模、工作界限，起到明确研究内容和限定标题及读者的作用。

(3) 结果或结论：简明扼要介绍研究所获得的结果或结论，有时还点明或暗示成果的价值或意义。

撰写引言时的注意事项：

(1) 不应与论文摘要雷同，或成为摘要的解释，不要注释基本理论，不要推导基本公式，不要介绍基本方法，不要重复教科书上已有的内容。

(2) 言简意赅，突出重点。在回顾前人所做的工作时，不要面面俱到，应找有代表性的，与本研究关系最密切的资料来阐述，避免写成文献综述。

(3) 客观、求实，不说废话、大话和套话。

常用的英语句型有：

(1) The experience research on... was carried out by...
(2) Recent experiments by... have suggested that...
(3) The previous work on... has indicated that...
(4) The paper is divided into five major sections as follows...

6. 论文主体

论文的主体部分篇幅大、内容多是主题思想的展开和论述。选材上要围绕主题段落划分，既要结构严谨，又要保证全文的整体性和连贯性。作者可根据需要在文章中加小标题，将主体内容分为几个部分进行论述。英文写作通常把每段的主题句 topic sentence 放在段落的第1句，全段围绕主题句论述，必要的例子和对实验数据的分析通常是不可缺少的。常用的句型有：

(1) The problem is chiefly concerned with the nature of... based on the study of...
(2) The core of the problem is the interaction of...
(3) It is described as follows...
(4) The present study was made with a view to show (demonstrate, determine)...
(5) Studies of these effects cover various aspects of...

7. 结论

结论是对研究或实验结果（仿真结果）的分析，应阐明作者的观点，指出争议的问题，得出最后的结论。常用的句型有：

(1) On the basis of..., the following conclusion can be made...

(2) From..., we now conclude (sum up) that...

(3) We have demonstrated in this paper...

(4) The results of the experiments (simulation) indicate (show)...

(5) Finally, a summary is given of...

8. 英文摘要示例

ABSTRACT: Aircraft engine fault diagnosis plays a crucial role in cost-effective operations of aircraft engines. However, designing an engine fault diagnostic system with the desired performance is a challenging task, because of several characteristics associated with aircraft engines. Geared towards achieving the highest possible performance of fault diagnosis, this paper explores strategies on improving diagnosis performance. Specifically, flight regime mapping and a two-level multiple classifier system is introduced as means to improve classification performance. By designing a real-world aircraft fault diagnostic system, it is demonstrated that the strategies adopted in this study are effective in improving the performance of aircraft engine fault diagnostic systems.

UNIT 20

ENGINE FUEL AND CONTROL SYSTEM

[1] The engine fuel and control system must supply fuel to the engine's fuel metering device under all conditions of ground and air operation. It must function properly at constantly changing altitudes and in any climate. Besides, the engine fuel and control system must be able to increase or decrease the power at will to obtain the thrust required for any operating condition. In turbine-powered aircraft, this control is provided by varying the flow of fuel to the combustion chambers.

[2] The quantity of fuel supplied must be adjusted automatically to correct for changes in ambient temperature or pressure. If the quantity of fuel becomes excessive in relation to mass airflow through the engine, the limiting temperature of the turbine blades can be exceeded, or it will produce compressor stall and a condition referred to as rich blowout. Rich blowout occurs when the amount of oxygen in the air supply is insufficient to support combustion and when the mixture is cooled below the combustion temperature by the excess fuel. The other extreme, lean flameout, occurs if the fuel quantity is reduced proportionally below the air quantity. The engine must operate through acceleration and deceleration without any fuel-control-related problems.

[3] The fuel system must deliver fuel to the combustion chambers not only in the right quantity, but also in the right condition for satisfactory combustion. The fuel nozzles form part of the fuel system and atomize or vaporize the fuel so that it ignites and burns efficiently. The fuel system must also supply fuel so that the engine can be easily started on the ground and in the air. This means that the fuel must be injected into the combustion chambers in a combustible condition during engine starting, and that combustion must be sustained while the engine is accelerating to its normal idling speed. Another critical condition to which the fuel system must respond occurs during a rapid acceleration. When the engine is accelerated, energy must be furnished to the turbine in excess of that necessary to maintain a constant rpm. However, if the fuel flow increases too rapidly, an over rich mixture can be produced, with the possibility of a rich blowout or compressor stall.

[4] **Fuel distribution system**: Here you can see a simplified engine fuel distribution system (shown in Fig. 20-1). This system usually starts directly behind the low pressure fuel shut-off valve. This valve feeds the fuel into the main fuel supply line, which runs from the wing to the engine accessory gearbox. The low pressure fuel pump increases the fuel pressure that comes from the tank boost pumps. The fuel from the low pressure pump then enters the oil cooler. This component has a dual function. The cold fuel cools the oil of the engine lubrication system and by this process the fuel is heated to a temperature above the freezing point of water. This prevents the ice particles coming from the fuel tanks, which can block the fuel filter. This is the reason why the fuel filter is located downstream of the oil cooler. It is needed to protect the following components in the engine fuel system.

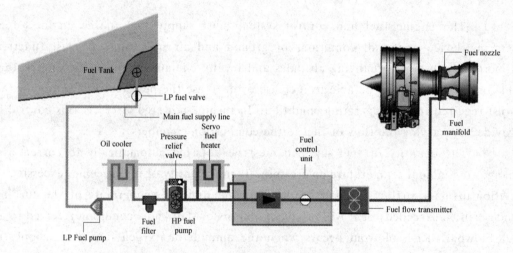

Fig. 20-1 Fuel Distribution System

[5] The next component, behind the fuel filter, is the high pressure fuel pump. This pump increases the fuel pressure to the high level needed for proper fuel vaporization in the combustion chamber. It is always equipped with a pressure relief valve which protects the components in the high pressure fuel system against overpressure. The fuel from the high pressure fuel pump then enters the fuel control unit. The fuel control unit meters the fuel that is needed for combustion. It is also responsible for supply and shut-off of fuel to the fuel nozzles at the combustion chamber.

[6] The fuel control unit needs some of the high pressure fuel as servo pressure to operate the internal control mechanisms. To be sure that this servo fuel is absolutely free of ice, some engines have an additional servo fuel heater. When the fuel leaves the fuel control unit, it has to pass through the fuel flow transmitter. The fuel flow transmitter measures the actual fuel flow and transmits signals to the cockpit for the fuel flow and fuel used indication.

[7] Note that all the components of the fuel distribution system, from the LP fuel pump to the fuel flow transmitter are in the area of the accessory gearbox at the engine. From the fuel flow transmitter, the fuel is then routed to the fuel manifold which distributes the fuel to the individual fuel nozzles on the combustion chamber.

[8] **Fuel control**: Gas turbine engine fuel controls can be divided into three basic groups: hydromechanical, hydromechanical/electronic and full authority digital engine (or electronics) control (FADEC). The hydromechanical/electronic fuel control is a hybrid of the two types of fuel control but can function solely as a hydromechanical control. In the dual mode, inputs and outputs are electronic, and fuel flow is set by servo motors. The third type, FADEC, uses electronic sensors for its inputs and controls fuel flow with electronic outputs. The FADEC-type control gives the electronic controller (computer) complete control. The computing section of the FADEC system depends completely on sensor inputs to the electronic engine control (EEC) to meter the fuel flow. The fuel metering device meters the fuel using only outputs from the EEC. Most turbine fuel controls are quickly going to the FADEC type of control.

[9] Electronic engine controls have allowed great increases in controlling the metered fuel flow to the engine. Engine fuel systems have become very accurate at providing the correct mixture of fuel and air to the engines. Gas turbine fuel controls have also greatly improved the ability to schedule (meter) the fuel correctly during all flight regimes. Improvements in electronics and the use of digital computers have enabled the aircraft and engines to be electronically interfaced together. By the use of electronic sensors and computer logic built in to electronic controls, the engines can be controlled with much more accuracy.

[10] Fuel cost and availability have also become factors in providing engines with fuel systems that are efficient and very precise in scheduling fuel flow to the engine. Many engines use an interactive system that senses engine parameters and feeds the information to the onboard computer (electronic engine control). The computer determines the amount of fuel needed and then sends a signal to the metering device. This signal sent to the metering device determines the correct amount of fuel needed by the engine. Electronic controls have become quite common with gas turbines and have increased the capabilities of the fuel system, making it less complicated for the technician and decreasing maintenance problems.

[11] Regardless of the type, all fuel controls accomplish essentially the same function. That function is to schedule the fuel flow to match the power required by the pilot. Some sense more engine variables than others. The fuel control can sense many different inputs, such as power lever position, engine rpm for each spool, compressor inlet pressure and temperature, burner pressure, compressor discharge pressure, and many more parameters as needed by the specific engine. These variables affect the amount of thrust that an engine produces for a given fuel flow. By sensing these

parameters, the fuel control has a clear picture of what is happening in the engine and can adjust fuel flow as needed. Each type of turbine engine has its own specific needs for fuel delivery and control.

[12] **Electronic Engine Control**: The electronic engine control (EEC) is the main component of the engine fuel and control system. On some engines it is also named the electronic control unit (ECU). The EEC uses analog and digital input data to calculate the engine fuel and control outputs to operate the engine. The EEC also gives digital data to other airplane systems. This data gives engine status.

[13] **FADEC system**: The FADEC system (shown in Fig. 20-2) has been developed to control fuel flow on most new turbine engine models. As the name indicates, in a FADEC system a digital computer has full authority over the engine control functions. The digital computer EEC is the heart of the FADEC system. The FADEC system must be fully redundant. This means, that each EEC has two independent computers. The two computers are called channel A and channel B. On some engines, the two computers are in individual housings. Each channel of the EEC receives individual signals from the aircraft and from the engine sensors. Signals from the engine to the EEC are either electric signals or pneumatic signals. For safety reasons all electrical signals are duplicated and transmitted individually to the two EEC channels.

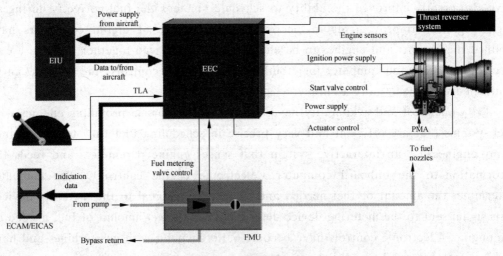

Fig. 20-2 FADEC System

[14] The data from the aircraft is used to improve the operation of the FADEC system, but the EEC can operate without it. During operation of the EEC the two channels are operating, but only one channel at a time has the authority to control the engine. This channel is called the channel in command. It generates the control signals. The other channel is in standby. The two channels are alternately in command. The switch-over from one channel to the other one is usually at the beginning of each engine start sequence.

[15] During engine operation the two channels A and B always collect the data from the engine and aircraft and they always communicate with each other via a cross channel data link. This internal communication is important to find out if the two channels are operating correctly and if all input data is correct. If one channel is faulty and unable to control, the command is taken over by the other channel. This change in command is automatic, based on the fault status of each channel. It is usually not visible to the operating personnel because the engine continues its operation.

[16] The second main component of the FADEC system is the fuel metering unit or FMU in short. This component is also named the hydromechanical unit or HMU on other engines. The terms FMU and HMU do not have a letter "C". This indicates that the FMU cannot control independently. It only receives orders from the EEC, to control the fuel metering valve.

[17] To operate correctly the electronic control unit needs the demand signal from the thrust lever in the cockpit. It also needs to know the engine speeds and all important air temperatures and air pressures in the engine and it certainly needs electrical power supply for its operation. This power supply either comes from the aircraft, or it can also come from a small permanent magnet alternator on the engine accessory gearbox. The EEC also needs a feedback signal about the opening condition of the fuel metering valve.

[18] The EEC can fulfill many tasks in addition to fuel metering and engine limit protection. It performs full power management and gives optimum thrust control for all operating conditions. It also controls other engine subsystems like the compressor stall protection system, the turbine and compressor clearance control system, thrust reverser system, the engine starting system and the engine indication system. With all the data that the EEC receives, it permanently monitors the engine operation and the important system components and gives fault messages to the centralized aircraft maintenance computer to indicate faulty components.

[19] On some aircraft there are so called engine interface units (EIUs) installed. These EIUs in the electronic compartment transmit the data between the FADEC system and the aircraft. There are only two direct connections between the aircraft and the electronic engine control. These signals, which are usually independent for safety reasons, are the thrust lever signal to the EEC and the engine indication data from the EEC. All the other signals between aircraft and EEC are routed via the engine interface unit like the ignition power supply and an electrical power supply for the EEC.

[20] **Engine control:** The engine control system supplies manual and automatic control inputs to operate the engine. It also supplies signals to other airplane systems that use engine control status. The components of engine control system are illustrated in Fig. 20-3, here we only introduce B737 series aircraft.

[21] **Thrust levers:** You use the thrust levers to supply the manual inputs to the engine control system. There are two thrust lever assemblies, one for each engine. A

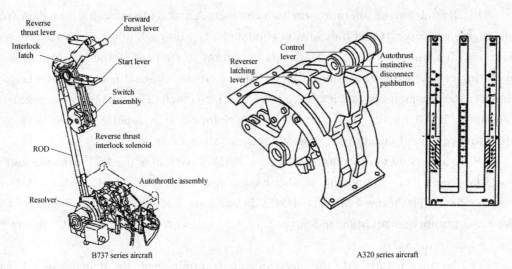

Fig. 20-3 Engine control lever

thrust lever assembly has many parts: forward thrust lever, reverse thrust lever, control link, crank, rod. For each engine, there is a forward thrust lever and a reverse thrust lever. The reverse thrust lever is on the forward thrust lever. For each engine, the thrust levers supply a thrust command signal to the electronic engine control (EEC) through the thrust lever resolver. Each thrust lever assembly connects mechanically to the resolver through an adjustable rod. An interlock latch prevents the forward thrust lever and the reverse thrust lever operating at the same time.

[22] The crank connects with the clutch pack and the resolver in the autothrottle assembly through the rod. The forward thrust lever and the crank are on the same shaft, but they move independently. The forward thrust lever holds the reverse thrust lever. The control link directly connects the reverse thrust lever and the crank. The control link moves up when you raise the reverse thrust lever. The control link moves down when you move the forward thrust lever forward.

[23] When you move the forward thrust lever, the position of the reverse thrust lever locks the control link onto the forward thrust lever. The force goes to the crank through the control link. When you move the reverse thrust lever, the force goes to the crank through the control link.

[24] **Thrust lever resolver**: There are two thrust lever resolver assemblies, one for each engine. Each thrust lever resolver assembly has two resolvers, one for EEC channel A and one for EEC channel B. The thrust lever resolvers change the mechanical forward and reverse thrust lever positions to analog thrust lever resolver angle (TRA) signals. These signals go to the EEC. The EEC uses these signals to control the engine.

[25] **Start levers**: There are two start levers, one for each engine. You use the engine start lever during an engine start. You also use it to shut down the engine. The start levers operate switches which supply signals to different aircraft and engine systems

and components. The start lever has two positions, IDLE and CUTOFF. A detent locks the lever in each position. You must pull the lever out to move it from one detent to the other. The lever connects mechanically to a brake which gives a friction force. Each start lever operates 6 switches. Two of the switches send signals to the EEC. Two of the switches interface with the engine ignition system. The other two switches send signals to valves in the engine fuel feed system.

[26] **Reverse thrust interlock solenoids**: The reverse thrust interlock solenoids energize to permit further movement of the reverse thrust levers during a T/R deploy operation. If the reverse thrust interlock solenoid does not energize, you cannot move the reverse thrust lever and increase reverse thrust. The solenoids energize when the T/R sleeves are 60% of travel to the full deploy position. Each EEC controls one of the solenoids.

[27] There are two reverse thrust interlock solenoids, one for each thrust lever assembly. They are a rotary solenoid type. Each reverse thrust interlock solenoid uses a rod to operate a latch. When you move a reverse thrust lever to the "DEPLOY" position, a contour on the brake mechanism catches the latch. This stops the rotation of the brake mechanism and limits the motion of the reverse thrust lever and the reverse thrust lever moves enough to operate switches to command the thrust reverser deployment. When the EEC energizes the reverse thrust interlock solenoid, the latch disengages. This permits the motion of the reverse thrust lever towards the full reverse thrust position.

NEW WORDS AND EXPRESSIONS

(1) rich blowout 富油熄火
(2) lean flameout 贫油熄火
(3) fuel nozzle 燃油喷嘴
(4) atomize['ætəmaɪz] 雾化
(5) idle['aɪdl] 慢车
(6) cooler['kuːlər] 散热器
(7) hydromechanical[ˌhaɪdrəʊmɪ'kænɪkəl] 液压机械的
(8) FADEC: full authority digital electronic control 全权限数字电子控制
(9) hybrid['haɪbrɪd] 混合物；混合
(10) EEC: engine electronic controller 发动机电子控制器
(11) spool[spuːl] 轴
(12) analog['ænəˌlɒg] 模拟
(13) duplicate['djuːplɪkeɪt] 复制
(14) channel['tʃænl] 通道；频道

(15) switch-over　　　　　　　　　　　　转变,转换
(16) FMU: fuel metering unit　　　　　　燃油计量装置
(17) HMU: hydraulic mechanical unit　　　液压机械装置
(18) EIU: engine interface unit　　　　　 发动机接口组件
(19) thrust lever　　　　　　　　　　　　油门杆,推力手柄
(20) resolver[rɪˈzɑlvər]　　　　　　　　 解算器
(21) autothrottle[ˌɔːtəʊˈθrɔtl]　　　　　 自动油门
(22) clutch[klʌtʃ]　　　　　　　　　　　离合器
(23) detent [dɪˈtent]　　　　　　　　　 制动器；(机械中的)棘爪掣,卡位
(24) interlock[ˌɪntərˈlɑːk]　　　　　　　联锁；互锁；联锁装置

EXERCISE

1. Translate the followings into Chinese.

(1) **Rich bloweout**. A condition in which the flame in a gas turbine engine goes out because of an overly rich fuel-air mixture. For fuel to burn in a turbine engine, it must be mixed with air in a ratio of between 8 and 18 parts of air for 1 part of fuel, by weight. If there is either more fuel or more air than this in the mixture, the flame will go out.

(2) **Lean flameout**. A condition in which the flame inside a gas turbine engine goes out because there is not enough fuel for the amount of air flowing through the engine. Turbine engine fuel controls are designed to measure the air flowing through the engine and adjust the fuel flow so a lean flameout, or lean die-out, cannot occur. A lean flameout is also called a lean die-out.

(3) **Idle thrust**. The thrust produced by an aircraft gas turbine engine when the power control lever is pulled back to the idle stop. The jet thrust obtained with the engine power control lever set at the stop for the least thrust position at which it can be placed.

(4) **Hydromechanical**. Any device that combines fluid pressures with mechanical actions to achieve a desired result. In a hydromechanical fuel control used for a turbine engine, hydraulic servos are used in conjuction with mechanical linkages.

(5) **EEC (electronic engine control)**. An electronic fuel control for a gas turbine engine. The EEC senses the power-lever angle (PLA), engine RPM, bleed valve, and variable stator vane position, and the various engine pressures and temperatures. It meters the correct amount of fuel to the nozzles for all flight conditions to prevent turbine overspeed and overtemperature.

(6) **Fuel control unit**. The fuel metering device for a gas turbine engine that senses

the variables of power lever position, pressure, speed, and temperature in the engine. The fuel control unit then meters the correct amount of fuel to be sent to the burners for the existing conditions.

(7) **Thrust reverser**. A device in the tail pipe of a gas turbine engine installed in an airplane that deflects some of the exhaust gases forward to produce a rearward thrust. This reverse thrust slows the aircraft and decreases its landing roll.

(8) **Clutch**. A component in a machine that connects or disconnects parts while they are moving. Clutches may be actuated by mechanical levers, hydraulic actuators, centrifugal force, or electromagnetism.

(9) **Detent**. A spring-loaded pin or tab that enters a hole or groove when the device to which it is attached is in a certain position. Detents are used on fuel selector valves to provide a positive means of identifying the position in which the valve is fully on and fully off.

(10) **Interlock**. A form of automatic safety control. An interlock in a device can prevent an action from occurring until the device is ready for the action. An interlock can be installed on the door of a compartment so the door cannot be opened when the equipment inside the compartment is operating. When the door is opened, the interlock switch automatically turns the power off and discharges any high-voltage capacitors inside the housing.

2. Answer the following questions in English.
(1) What is rich blowout? And what is lean flamout?
(2) What is the function of engine fuel system?
(3) What are the types of engine fuel control system?
(4) What are the components of engine control system?
(5) How to control the gas turbine engine by using thrust lever? Taking Airbus aircraft or Boeing aircraft as an example.

SUPPLEMENTARY READING

[1] **Turbine engine instruments**: Engine instruments that indicate oil pressure, oil temperature, engine speed, exhaust gas temperature, and fuel flow are common to both turbine and reciprocating engines. However, there are some instruments that are unique to turbine engines. These instruments provide indications of engine pressure ratio, turbine discharge pressure, and torque. In addition, most gas turbine engines have multiple temperature-sensing instruments, called thermocouples, which provide pilots with temperature readings in and around the turbine section.

[2] **Engine pressure ratio (EPR) gauge**: This gauge is used to indicate the power output of a turbojet/turbofan engine. EPR is the ratio of turbine discharge to

compressor inlet pressure. Pressure measurements are recorded by probes installed in the engine inlet and at the exhaust. Once collected, the data is sent to a differential pressure transducer, which is indicated on a flight deck EPR gauge. EPR system design automatically compensates for the effects of airspeed and altitude. Changes in ambient temperature require a correction be applied to EPR indications to provide accurate engine power settings.

[3] **Exhaust gas temperature (EGT) gauge**: A limiting factor in a gas turbine engine is the temperature of the turbine section. The temperature of a turbine section must be monitored closely to prevent overheating the turbine blades and other exhaust section components. One common way of monitoring the temperature of a turbine section is with an EGT gauge. EGT is an engine operating limit used to monitor overall engine operating conditions. Variations of EGT systems bear different names based on the location of the temperature sensors. Common turbine temperature sensing gauges include the turbine inlet temperature (TIT) gauge, turbine outlet temperature (TOT) gauge, interstage turbine temperature (ITT) gauge, and turbine gas temperature (TGT) gauge.

[4] **Torquemeter**: Turboprop/turboshaft engine power output is measured by the torquemeter. Torque is a twisting force applied to a shaft. The torquemeter measures power applied to the shaft. Turboprop and turboshaft engines are designed to produce torque for driving a propeller. Torquemeters are calibrated in percentage units, foot-pounds, or psi.

[5] **N1 indicator**: N1 represents the rotational speed of the low pressure compressor and is presented on the indicator as a percentage of design rpm. After start, the speed of the low pressure compressor is governed by the N1 turbine wheel. The N1 turbine wheel is connected to the low pressure compressor through a concentric shaft.

[6] **N2 indicator**: N2 represents the rotational speed of the high pressure compressor and is presented on the indicator as a percentage of design rpm. The high pressure compressor is governed by the N2 turbine wheel. The N2 turbine wheel is connected to the high pressure compressor through a concentric shaft.

[7] **Tachometer**: Gas turbine engine speeds are measured by the engine's rpm, which are also the compressor/turbine combination rpm of each rotating spool. Most turbofan engines have two or more spools, compressor, and turbine sections that turn independently at different speeds. Tachometers are usually calibrated in percent rpm so that various types of engines can be operated on the same basis of comparison. Also, turbine speeds are generally very high, and the large numbers of rpm would make it very confusing. Turbofan engines with two spools or separate shafts, high pressure and low pressure spools, are generally referred to as N1 and N2, with each having their own indicator. The main purpose of the tachometer is to be able to monitor rpm under normal conditions, during an engine start, and to indicate an overspeed condition, if one occurs.

[8] **Fuel-flow indicator:** Fuel-flow instruments indicate the fuel flow in pounds per hour (lb/h) from the engine fuel control. Fuel flow in turbine aircraft is measured in lb/h instead of gallons, because the fuel weight is a major factor in the aerodynamics of large turbine aircraft. Fuel flow is of interest in monitoring fuel consumption and checking engine performance.

[9] **Engine oil pressure indicator:** To guard against engine failure resulting from inadequate lubrication and cooling of the various engine parts, the oil supply to critical areas must be monitored. The oil pressure indicator usually shows the engine oil pump discharge pressure.

[10] **Engine oil temperature indicator:** The ability of the engine oil to lubricate and cool depends on the temperature of the oil, as well as the amount of oil supplied to the critical areas. An oil inlet temperature indicator frequently is provided to show the temperature of the oil as it enters the oil pressure pump. Oil inlet temperature is also an indication of proper operation of the engine oil cooler.

TOPIC 20：英文电子邮件写作规范

英文电子邮件的基本要素是：主题、称谓、正文、结尾用语及署名。

1. 主题（Subject）

在打开邮箱阅读邮件的时候，第一眼看到的就是邮件主题，因此主题框的内容应简明概括邮件的内容，短的可以是一个单词，如 greetings；长的可以是一个名词性短语，也可以是完整句，但长度一般不超过 35 个字母。

> 正确：Supplier training
> 错误：professional trainees from sister company should abide by the rule of local company（太长）

句首单词和专有名词首字母大写，在比较正规的格式中，将每个单词的首字母大写（介词、冠词、连词除外）。

> 正确：New E-mail Address Notification；Detailed Calculation
> 错误：detailed calculation

根据信的内容是否重要和紧急，还可以在开头加上 URGENT 或者 FYI（For Your Information，供参考），如：URGENT：Submit your report today！

2. 称谓（Salutation）

英文电子邮件一般使用非正式的文体，正文前的称谓通常为其名字，但对长辈或者上级最好使用头衔加上姓，如 Mr. Smith。

如果是第一次给对方写信或者不知道对方姓名时，称谓最好用"Dear + 全名"，这样感觉会比较正式，如"Dear President"。英国人习惯在称谓后加"，"，美国人习惯在称谓后加"："，有时也可以不加任何标点，视具体情况而定。

称谓和正文之间、段落之间、正文和信尾客套话之间一般空一行，开头不需要空格。

3. 正文（Body of Letter）

在书写正文时，把最重要的事情写在正文最前面或者邮件内容较长时写在第一段。为了让收件人阅读邮件时比较舒服，应该注意邮件正文结构的美感，邮件段落最好控制在两三段之内。如果一封电子邮件涉及多个信息点，可以采用分条目的方法，如符号、编号等来使邮件想要表达的内容层次清晰。邮件内容应注意单词的拼写、大小写、标点符号、语法等。

大小写要注意，非必要不要整个单词都是大写。如果要强调的话，可以用底线、斜体字、粗体。

> 错误：MUST change to OS immediately.
> 正确：*Must* change to OS immediately.

4. 结尾用语（Complimentary Close）

结尾用语在正文之后。一般结尾用语中只有第一个单词首字母大写，而剩余单词都小写,与前面所讲的称谓不同。

(1) Very Formal 非常正规(如写给政府官员)

Respectfully yours,Yours respectfully

(2) Formal 正规的(客户公司之间)

Very truly yours,Yours very truly,Yours truly

(3) Less Formal 不太正规(如客户)

Sincerely yours,Yours sincerely,Sincerely

(4) Informal 非正规(朋友、同事)

Regards,Warm regards,My best,Best wishes,Thanks,See you next week!

5. 署名(Signature)

在正文最后需要署名,可以写全名,也可以只写名字,需要辨明性别时可以在姓名后面注明(Mr./Ms.)。对于中国人的姓名而言,为了区分姓和名,可以把姓的字母全部大写,例如 David TONG。如果写信人代表的是一个组织或部门,应在名字下一行写上自己的职位和所属部门。

6. 邮件中常用的词汇

(1) 附件(attachment):I am attaching the report.

(2) 转发(forward):I've forwarded your E-mail to John.

(3) 保持联系:Stay/keep in touch.

(4) 等你好消息:Awaiting your good news.

(5) 期待你的早日回复:Looking forward to your early reply.

(6) 请对方告知决定:Please let me know your decision as soon as possible.

(7) 写信给别人时开头常用:How have you been recently?

(8) 回复邮件时开头常用:I am happy to receive your E-mail.

7. 实用邮件举例

(1) 工作求职:包括申请职位、介绍自己、期待回复、表达谢意。

> Dear sir/madam
>
> I am writing to apply for the position of... as posted in your website.
>
> Introduce yourself...
>
> Please find attached a copy of my resume for your review.
>
> I am looking forward to you. Thank you very much for your time and consideration.
>
> Yours sincerely
>
> David TONG

(2) 表达谢意:表达自己对收件人真挚的感激之情。

> Dear Alice,
>
> Thanks so much for the lovely dinner last night. It was so thoughtful of you. I would like to invite both of you to my house when you are available.
>
> Best regards.
>
> David

(3) 咨询了解：目的是想要得到关于某人某事的一些信息，内容大致包括：
- 告知对方你是如何得知其信息的：报纸，网站等；
- 说明自己想要了解的东西，如果咨询信息较多时，最好列举出来；
- 末尾用一句话总结这次咨询：

I am writing to request information about sth.

I would like to request a copy of sth.

参 考 文 献

[1] A320 Aircraft Maintenance Manual. Airbus Company[Z]. 2003.
[2] B787 Aircraft Maintenance Manual. Boeing Company[Z]. 2014.
[3] B737NG Aircraft Maintenance Manual. Boeing Company[Z]. 2014.
[4] CRANE D. Dictionary of Aeronautical Terms [M]. Newcastle: Aviation Supplies & Academics, Inc. 1997.
[5] BRIEGER N, COMFORT J. 技术英语[M]. 北京:清华大学出版社,1999.
[6] U.S. Department of Transportation Federal Aviation Administration Flight Standards Service. Airframe and Powerplant Mechanics Airframe Handbook [M]. U.S. Summit Aviation, Inc. 2000.
[7] U.S. Department of Transportation Federal Aviation Administration Flight Standards Service. Pilot's Handbook of Aeronautical Knowledge [Z]. U.S. Summit Aviation, Inc. 2003.
[8] U.S. Department of Transportation. FAA Flight Standards Service. Aviation Maintenance Technician Handbook-Airframe (Volume 1)[Z]. 2012.
[9] U.S. Department of Transportation. FAA Flight Standards Service. Aviation Maintenance Technician Handbook-Airframe (Volume 2)[Z]. 2012.
[10] MASON W H. Modern Aircraft Design Techniques[M]. Blacksburg: Department of Aerospace and Ocean Engineering Virginia Tech, 2003.
[11] 冯庆华. 实用翻译教程英汉互译[M]. 上海:上海外语教育出版社,2000.
[12] 规范:ASD-STE100 简化技术英语[Z].
[13] 顾忆华,许珂,王立国. 大学应用英语写作[M]. 北京:中国人民大学出版社,2018.
[14] 胡静. 航空器制造工程专业英语[M]. 西安:西安电子科技大学出版社,2015.
[15] 李学平. 科技翻译与英语学习[M]. 天津:南开大学出版社,2005.
[16] 刘振海. 中英文科技论文写作[M]. 北京:高等教育出版社,2012.
[17] 芬克斯坦. 科技写作教程[M]. 北京:清华大学出版社,2011.
[18] 王宏文. 自动化专业英语教程[M]. 北京:机械工业出版社,2015.
[19] 许卉艳. 科技英汉互译教程[M]. 北京:知识产权出版社,2015.
[20] 张铁纯. 涡轮发动机飞机结构与系统(ME-TA)[M]. 2版. 北京:清华大学出版社,2017.